INFORMATION STORAGE AND MANAGEMENT

Pankaj Sharma
(Astt. Prof.)
Department of Information Technology
ABES, Ghaziabad

S.K. KATARIA & SONS®
Publisher of Engineering Books
4885/109, Prakash Mahal, Dr. Subhash Bhargav Lane,
Opposite Delhi Medical Association, Daryaganj, New Delhi–110002 (INDIA)
Phone: +91-11-43551243; *Mobile:* +91-9871775858, 7678425209
e-mail: katariabook@gmail.com, katariabooks@yahoo.com
Website: www.skkatariaandsons.com
(e-books, subscriptions and bulk buying also available on our website)

Published by:

S.K. Kataria & Sons®

4885/109, Prakash Mahal, Dr. Subhash Bhargav Lane,
Opposite Delhi Medical Association, Daryaganj, New Delhi–110002
Phone: +91-11-43551243; Mobile: +91-9871775858, 7678425209
e-mail: katariabook@gmail.com, katariabooks@yahoo.com

First Edition: August, 2011; Second Edition: July, 2012, Reprint: 2026

Price: ₹ 295/-

Laser Typesetting By :
PLUS COMPUTERS, Shahdara, Delhi-110093.

Printed at :
Arora Printing Press, Delhi–110051

PREFACE

Information Storage and Management is a core subject for undergraduate and Post graduate students in Computer Science Engineering (CSE), Information Technology Streams and Master of Computer Applications (MCA). This book is intended to serve as a textbook in accordance with the syllabus of Information Storage and Management offered by various universities in India as well as abroad.

In this book, a significant effort has been made to find simple ways to develop theoretical as well as Practical concepts of Rapid Application Development. A very strong emphasis, has been given on Practical examples. Neat and clean diagrams have been used for explanation.

Every effort has been made to make the concepts simple and comprehensive. This book is divided into eight chapters. The organization of these chapters are:

Chapter 1	Gives the Introduction to Various Storage Technologies
Chapter 2	Describes the Storage System Architecture
Chapter 3	Introduces the Intelligent Disk Sub System
Chapter 4	Explains the Concept of Networked Storage
Chapter 5	Describes the Concept of Hybrid Storage Solution
Chapter 6	Introduces the Case Study as Standard Application Framework for EPICS
Chapter 7	Gives the Introduction to Key Management Metrices
Chapter 8	Explains the Concept of Information Storage as Cloud.

Pankaj Sharma
ABES, Ghaziabad

ACKNOWLEDGEMENT

I get this opportunity to express my gratitude and thanks to the respected Director (administration) of ABES Engg. College, **Dr. S.K. Kansal** for his valuable technical suggestion and constant encouragement, without which this book would not have to come into existence.

I express kind gratitude of **Prof. R.G. Mendiratta**, Director (Academic) ABES Engg. College, Ghaziabad.

I wish to thank **Prof. C.L. Mittal** (Dean UG) ABES Engg. College, Gzb. I also like to remember the motivation initiated by **Shri M.D Punia** my father-in-Law, (Lead District Manager, PNB, Karnal).

I am grateful to **Prof. V.K. Srivastava,** Director of ABES, IT, Ghaziabad, for helping me through his insightful suggestions and construction inputs.

I wish to thank **Prof. Pradeep Sharma**, Head of Department of Information Technology, Department of ABES Engg. College, Ghaziabad, for their constant co-operation, directly inspiration and encouragement.

I like to remember the motivation initiated by **Prof. D. Pandey** (HOD CSE) and Prof. Sunil Sharma (AHODCSE), ABES, Engg. College. I am also thankful to Prof. K.P. Jayant (HODMCA) for his inspiration.

I thank to my elder brother **Mr. Sanjay Sharma** (KIET), Ghaziabad for inspiration and encouragement.

I thank to **Mr. Aswin Perti,** Lecturer in IT Department for his technical suggestions.

I am grateful to my younger Brother **Sandeep** and daughter **Prisha** for inspiring me for this project.

I deeply express my heartful thanks to the **Mr. Sanjeev Kataria** for **S.K. Kataria & Sons** for publishing this book in such a beautiful getup and well in time.

Lastly I am grateful to Mr. Manoj Tyagi and his staff from Plus Computer's for thier hardwork.

Pankaj Sharma

Dedicated This book to

My Parents, Wife Priyanka, Daughter Prisha, Pihu

and

Readers of this book

CONTENTS

CHAPTER

1

INTRODUCTION TO STORAGE TECHNOLOGIES

> **Inside this Chapter**
> - Introduction
> - Types of Storage Technology
> - Data proliferation
> - Evolution of various Storage Technologies
> - Overview of storage infrastructure components
> - Information Lifecycle Management
> - Data categorization

1.0. INTRODUCTION

There are many types of storage media:

- Flash, which has become a cheap form of fast storage, especially in consumer products.
- Optical storage, which comes in the form of CDs, DVDs and BluRay etc. These are slow for data access, but still very useful for archives and movies.
- Magnetic Tape backup systems, which are still in use in corporate IT centers, but are slow, and aren't good for random access. Random access refers to the ability to (effectively) access any piece of data by its address (e.g. block number on a hard disk – see below) instantly.
- Magnetic or hard disks, which are discussed at length below, are a ubiquitous form of high-volume, high-speed, random-access storage.
- A solid-state drive (SSD) is a data storage device that uses solid-state memory to store persistent data. Unlike flash-based memory cards, an SSD emulates a hard disk drive, thus easily replacing it in most applications. An SSD using SRAM or DRAM (instead of flash memory) is often called a RAM drive.

The advantage over (magnetic) disk drives is speed, but the cost per gigabyte is 4 to 5 times that of disk drives, and at the moment the amount of storage is much less per unit.

1.1. TYPES OF STORAGE TECHNOLOGY

Data, whether structured or unstructured, does not fulfill any purpose for individuals or businesses unless it is presented in a meaningful form. Businesses need to analyze data for it to be of value. *Information* is the intelligence and knowledge derived from data.

Businesses analyze raw data in order to identify meaningful trends. On the basis of these trends, a company can plan or modify its strategy. For example, a retailer identifies customers' preferred products and brand names by analyzing their purchase patterns and maintaining an inventory of those products. Effective data analysis not only extends its benefits to existing businesses, but also creates the potential for new business opportunities by using the information in creative ways. Job portal is an example. In order to reach a wider set of prospective employers, job seekers post their résumés on various websites offering job search facilities. These websites collect the résumés and post them on centrally accessible locations for prospective employers. In addition, companies post available positions on job search sites. Job-matching software matches keywords from.

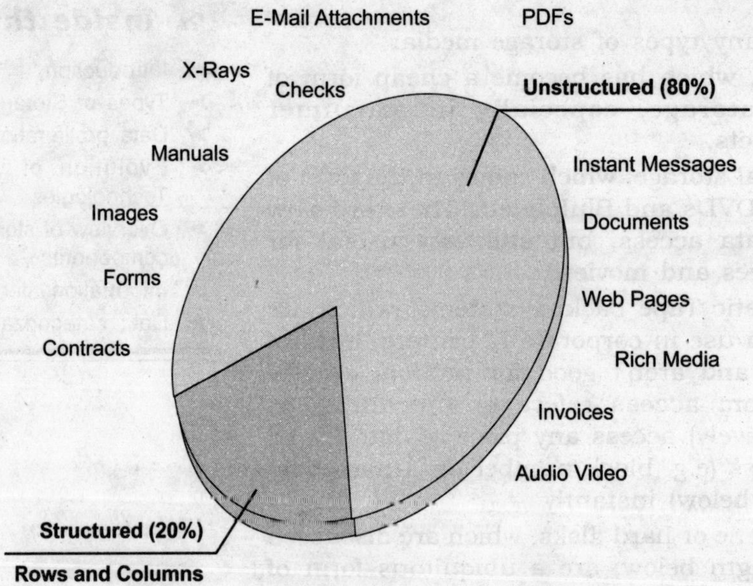

Fig. 1.1.

Before the advent of computers, the procedures and methods adopted for data creation and sharing were limited to fewer forms, such as paper and film. Today, the same data can be converted into more convenient forms such as an e-mail message, an e-book, a bitmapped image, or a digital movie. This data can be generated using a computer and stored in strings of 0s and 1s, as shown in Fig. 1.2. Data in this form is called *digital data* and is accessible by the user only after it is processed by a computer.

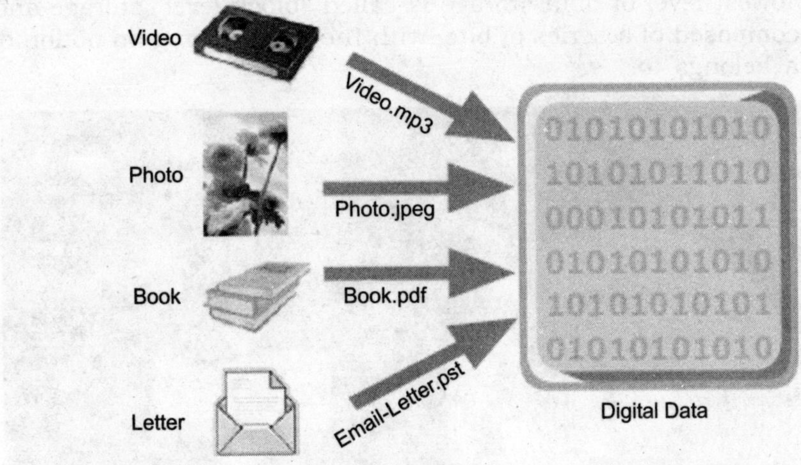

Fig. 1.2.

These types of storage can either be static or removable thanks to the ubiquity of USB and firewire. In this paper we will be exclusively talking about magnetic disk (as in hard-drive) storage.

1.1.1. Hard Disk

Hard disks are composed of:

- Multiple spinning magnetic platters that contain the magnetically encoded data. The platters spin around an axle called a "spindle", and sometimes a single drive (like the one shown) can be referred to as such.

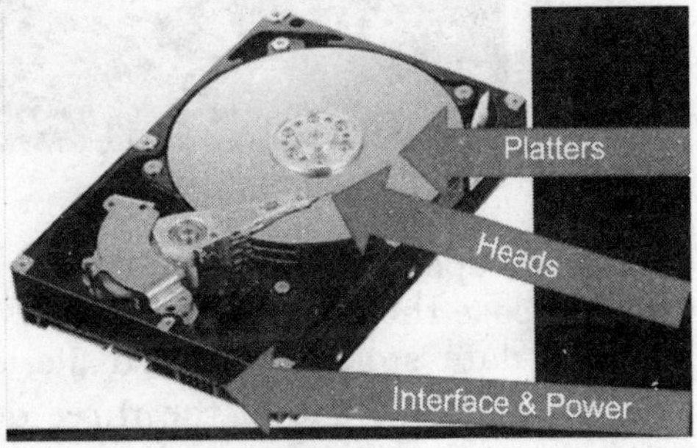

Fig. 1.3.

- Read/write heads that float above the surface of the platters, usually 2 (top and bottom) for each platter. The heads all move in unison.
- A controler board that drives the heads and can convert I/O requests (commands) into head movement and read/write operations.
- An interface that will be joined to a host adapter board, which for an internal (to a computer) drive, is a separate unit.
- The drive will need power which, for an internal drive, is usually supplied by the computer's own power supply.

The disk drive is covered and hermetically sealed, since heads are designed to float above the disk platter with less than a micron of space.

Each surface of the disk platter is divided into areas as shown. In the context of data, the word "block" can have many different meanings, but in this context a block is the smallest unit of data that is read and written. A block is also called (more formally) a "track sector" or "simply sector" as shown by (C). A series of sectors makes up a track (A), and there are several tracks on a surface.

At the lowest level when a block (or segment) is being read or written, there are several things that identify where that block goes: The surface (identified by the head number), the track that the head should move to, and the sector that should be read or written. This lowest level of data storage is called "block-level" storage and implies that the data is composed of a series of bits, with the drive having no notion of format, or what the data belongs to.

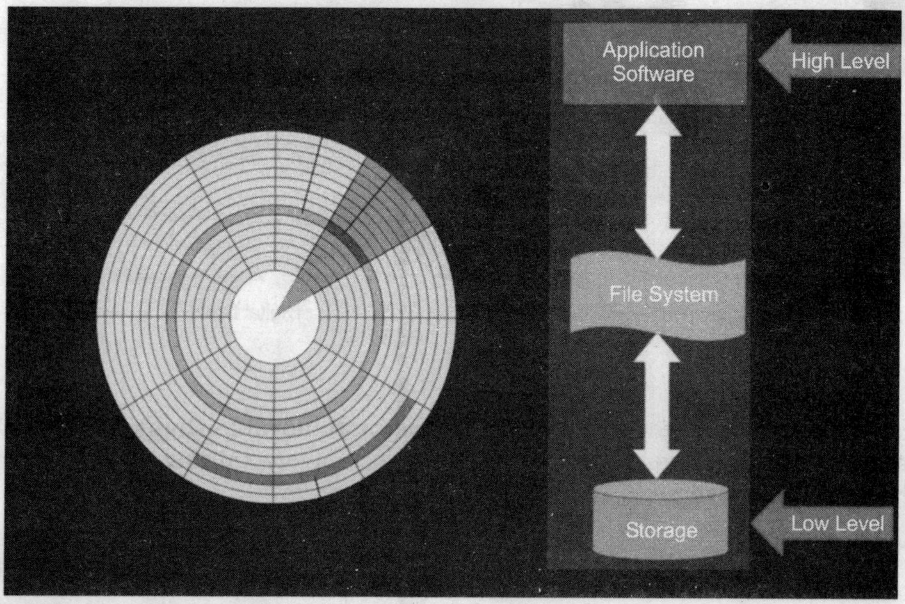

Fig. 1.4.

In the operating system (OS), there will be device drivers, file systems and applications that impose the meaning on, and keep tabs on the individual blocks of data. This high level of data storage is called file-level storage.

Seek time is one of the three delays associated with reading or writing data on a disk drive. The others are the rotational delay of the disk, and transfer time. Their sum is the access time. In order to read or write data in a particular sector, the head of the disk needs to be physically moved to the correct place. This process is known as seeking, and the time it takes for the head to move to the right place is the seek time. Seek time for a given disk varies depending on how far the head's destination is from its origin at the time of each read or write; usually one discusses a disk's average seek time.

1.1.2. Host Controller Card

For disk drives that are to be installed internally to a computer (such as a server), the interface from the disk will be cabled to a Host Controller Card (or simply controller). Depending on the type of controller, this can usually accommodate multiple disk drives. You may also be able to plug in external devices to an external-facing interface.

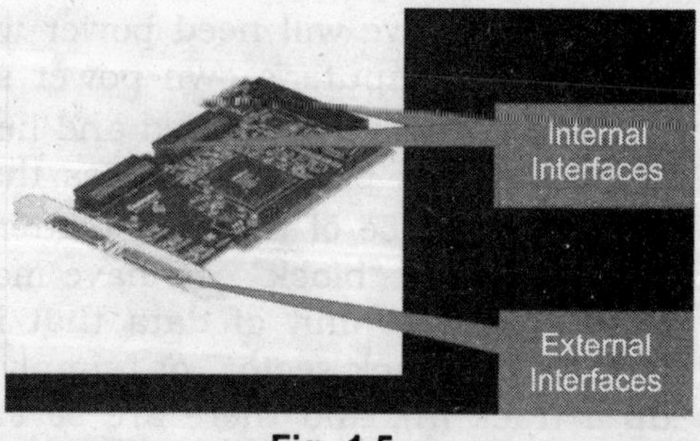

Fig. 1.5.

The controller will usually plug straight into a slot on the computer motherboard and draw its power from there. This will also let the CPU talk to the host adapter and disks through the system "bus". Sometimes you may find the controller actually integrated into the computer's motherboard.

1.1.3. Types of Disk Interfaces

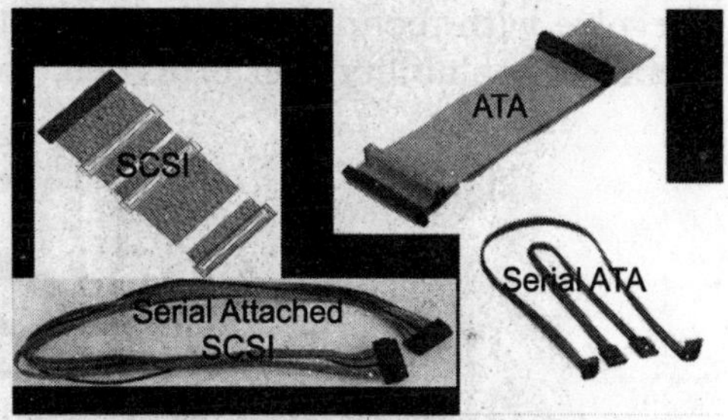

Fig. 1.6.

The predominant interfaces for disk drives are:

- **Advanced Technology Attachment (ATA):** This is more common in home computing, and can support different kinds of devices, such as hard drives and DVD burners. There is a restriction of 2 devices per cable. This is also known as a parallel ATA, since there is a wire for each bit of information in the interface cable, making it very wide. The disk drives that are attached to an ATA host adapter are usually called IDE drives (Integrated Drive Electronics).
- **Serial ATA (SATA):** Is an enhancement to ATA, which allows for changing drives without shutting down (hot swap), faster transfer speeds, and thinner cabling.

1.2. DATA PROLIFERATION

Data proliferation refers to the prodigious amount of data, structured and unstructured, that businesses and governments continue to generate at an unprecedented rate and the usability problems that result from attempting to store and manage that data. While originally pertaining to problems associated with paper documentation, data proliferation has become a major problem in primary and secondary data storage on computers.

While digital storage has become cheaper, the associated costs, from raw power to maintenance and from metadata to search engines, have not kept up with the proliferation of data. Although the power required to maintain a unit of data has fallen, the cost of facilities which house the digital storage has tended to rise.

The problem of data proliferation is affecting all areas of commerce as the result of the availability of relatively inexpensive data storage devices. This has made it very easy to dump data into secondary storage immediately after its window of usability has passed. This masks problems that could gravely affect the profitability of businesses and the efficient functioning of health services, police and security forces, local and national governments, and many other types of organization.

The importance, dependency, and volume of information for the business world also continue to grow at astounding rates. Businesses depend on fast and reliable access to information critical to their success. Some of the business applications that process information include airline reservations, telephone billing systems, e-commerce, ATMs, product designs, inventory management, e-mail archives, Web portals, patient records, credit cards, life sciences, and global capital markets.

The increasing criticality of information to the businesses has amplified the challenges in protecting and managing the data. The volume of data that business must manage has driven strategies to classify data according to its value and create rules for the treatment of this data over its lifecycle. These strategies not only provide financial and regulatory benefits at the business level, but also manageability benefits at operational levels to the organization.

Data centers now view information storage as one of their core elements, along with applications, databases, operating systems, and networks. Storage technology continues to evolve with technical advancements offering increasingly higher levels of availability, security, scalability, performance, integrity, capacity, and manageability.

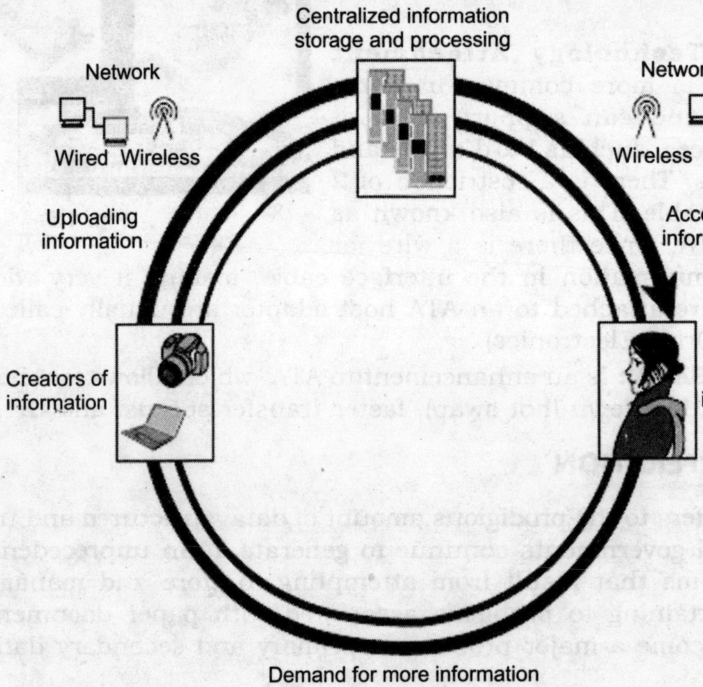

Fig. 1.7. Data proliferation process

1.3. EVOLUTION OF VARIOUS STORAGE TECHNOLOGIES

Storage systems are built by taking the basic capability of a storage device, such as the hard disk drive, and adding layers of hardware and software to obtain a highly reliable, high-performance, and easily managed system. We explain in this paper how storage systems have evolved over five decades to meet changing customer needs. First, we briefly trace the development of the control unit, RAID (redundant array of independent disks) technologies, copy services, and basic storage management technologies. Then, we describe how the emergence of low-cost local area data networking has allowed the development of network-attached storage (NAS) and storage area network (SAN) technologies, and we explain how block virtualization and SAN file systems are necessary to fully reap the benefits of these technologies. We also discuss how the recent trend in storage systems toward managing complexity, easeof-use, and lowering the total cost of ownership has led to the development of autonomic storage. We conclude with our assessment of the current state-of-the-art by presenting a set of challenges driving research and development efforts in storage systems.

The first data storage device was introduced by IBM in 1956. Since then there has been remarkable progress in hard disk drive (HDD) technology, and this has provided the fertile ground on which the entire industry of storage systems has been built. *Storage systems* are built by taking the raw storage capability of a storage device such as the HDD and by adding layers of hardware and software in order to obtain a system that is highly reliable, has high performance, and is easily manageable. Storage systems are sometimes referred to as *storage subsystems* or *storage devices* (although *device* is better used to describe the raw storage component or an elementary storage system).

Originally, the storage system was just the HDD, but over time storage systems have developed to include advanced technologies that add considerable value to the HDD. Storage systems have evolved to support a variety of added services, as well as connectivity and interface alternatives. It is for this reason that file systems and storage management systems are often considered parts of a storage system and thus will be briefly treated in this chapter. To understand the evolution of storage systems, it is important to observe the evolution of the HDD.

The areal density of the HDD has improved by seven orders of magnitude, and this has resulted in a reduction of the floor space taken by the corresponding storage systems also by about seven orders of magnitude. Figure 1.5. plots the HDD areal density (on the left) and the price of various storage devices (on the right) since 1980. HDD prices have decreased by about five orders of magnitude since 1980, while the cost of storage systems has fallen about of magnitude in the same period.1 The sharper fall in the price of the HDD implies that the cost of the raw HDD accounts for a progressively smaller fraction of the total cost of the storage system—we will return to this crucial observation later. Although Moore's law tells us that the number of transistors per unit area of silicon doubles every 1.5 years, we see from Fig. 1.5. That the number of bits stored per unit of HDD media is doubling about every year! But more important than improvements in device density or cost have been the new applications that have been enabled by these advances. On the time line in Fig. 1.5, two milestones stand out. In 1996, digital storage became more cost effective for storing data than paper, and, in 1998, we reached the point where film used in medical radiology could be economically supplanted by electronic means. Another important milestone, this one in the consumer market, was reached several years ago when it became cost effective to store video content using digital storage systems. Soon after we saw the emergence of HDD-based set-top-box devices (sometimes called personal video recorders) that offered improved management of entertainment video in the home.

The proliferation of departmental servers in an enterprise resulted in unprotected, unmanaged, fragmented islands of information and increased operating cost. Originally, there were very limited policies and processes for managing these servers and the data created. To overcome these challenges, storage technology evolved from non-intelligent internal storage to intelligent networked storage. Highlights of this technology evolution include:

Storage technology and architecture continues to evolve, which enables organizations to consolidate, protect, optimize, and leverage their data to achieve the highest return on information assets.

- **Redundant Array of Independent Disks (RAID):** This technology was developed to address the cost, performance, and availability requirements of data. It continues to evolve today and is used in all storage architectures such as DAS, SAN, and so on.
- **Direct-attached storage (DAS):** This type of storage connects directly to a server (host) or a group of servers in a cluster. Storage can be either internal or external to the server. External DAS alleviated the challenges of limited internal storage capacity.
- **Storage area network (SAN):** This is a dedicated, high-performance *Fibre Channel (FC)* network to facilitate *block-level* communication between servers and storage.

Storage is partitioned and assigned to a server for accessing its data. SAN offers scalability, availability, performance, and cost benefits compared to DAS.

- **Network-attached storage (NAS):** This is dedicated storage for *file serving* applications. Unlike a SAN, it connects to an existing communication network (LAN) and provides file access to heterogeneous clients. Because it is purposely built for providing storage to file server applications, it offers higher scalability, availability, performance, and cost benefits compared to general purpose file servers.
- **Internet Protocol SAN (IP-SAN):** One of the latest evolutions in storage architecture, IP-SAN is a convergence of technologies used in SAN and NAS. IP-SAN provides block-level communication across a local or wide area network (LAN or WAN), resulting in greater consolidation and availability of data.

1.4. OVERVIEW OF STORAGE INFRASTRUCTURE COMPONENTS

Storage systems come of age from components to systems: It has long been recognized that the disk drive alone cannot provide the range of storage capabilities required by enterprise systems. The first storage devices were directly controlled by the CPU. Although some advances did take place in the interim, System/360* in 1964 was the first offering of advanced functions in an external storage control unit. The key advantage of a control unit (or controller) was that the I/O commands from the CPU (sometimes called the host) were independently translated into the specific commands necessary to operate the HDD (sometimes called the direct access storage device, or DASD), and so the HDD device itself could be managed independently and asynchronously from the CPU.

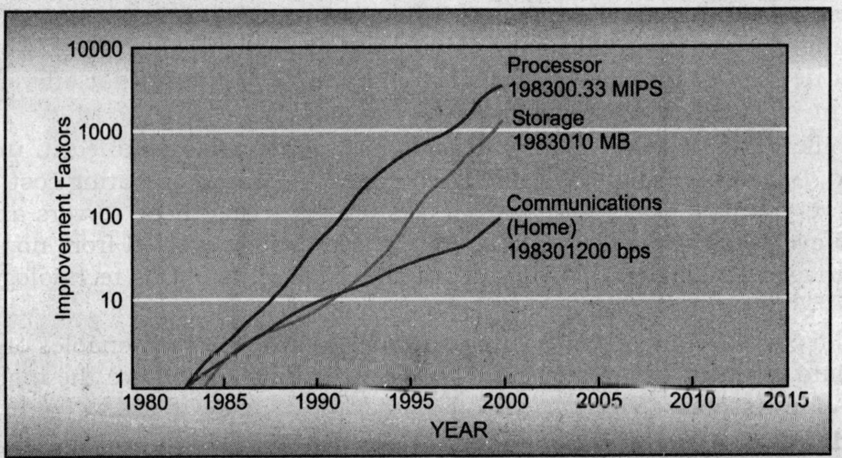

Fig. 1.8. Improvement factors for PC technologies: since 1990 storage technology has outpacked both processor and communication technologies

The control unit included buffering that allowed. The CPU and HDD operations to overlap. Over time there was a greater emphasis on availability, and controllers began to support multiple data paths from CPUs to storage controllers. The IBM 3990 Model 3 storage controller consisted of two clusters on separate power and service boundaries and included both a large cache and nonvolatile storage (NVS) memories. The caches were used to improve read response time, whereas the NVS was used to provide a fast write and dual copy capability. In addition, it was possible to perform maintenance on the cluster, cache, or NVS under normal operation. Storage systems leapt further ahead in the early 1990s when RAID (redundant array of independent disks) technology was introduced. RAID allowed the coordination of multiple HDD devices so as to provide higher levels of reliability and performance than could be provided by a single drive.

Five core elements are essential for the basic functionality of a data center:

- **Application:** An application is a computer program that provides the logic for computing operations. Applications, such as an order processing system, can be layered on a database, which in turn uses operating system services to perform read/write operations to storage devices.
- **Database:** More commonly, a database management system (DBMS) provides a structured way to store data in logically organized tables that are interrelated. A DBMS optimizes the storage and retrieval of data.
- **Server and operating system:** A computing platform that runs applications and databases.
- **Network:** A data path that facilitates communication between clients and servers or between servers and storage.
- **Storage array:** A device that stores data persistently for subsequent use.

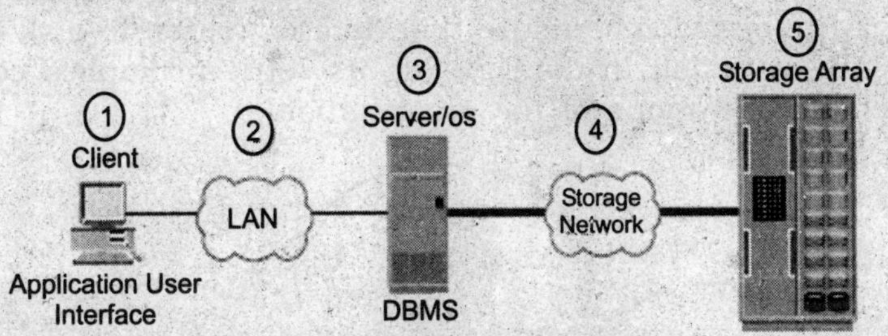

Fig. 1.9.

A customer places an order through the AUI of the order processing application software located on the client computer.

The clients connects to the server over the LAN and accesses the DBMS located on the server to update the relevant information such as the customer name, address, payment method, products ordered, and quantity ordered.

The DBMS uses the server operating system to read and write this data to the database located on physical disks in the storage array.

The storage network provides the communication link between the server and the storage array and transports the read or write commands between them.

The storage array, after receiving the read or write commands from the server, performs the necessary operations to store the data on physical disks.

1.5. INFORMATION LIFECYCLE MANAGEMENT

Information Lifecycle Management refers to a wide-ranging set of strategies for administering storage systems on computing devices. Specifically, four categories of storage strategies may be considered under the auspices of ILM.

Information lifecycle management (ILM) is a proactive strategy that enables an IT organization to effectively manage the data throughout its lifecycle, based on predefined business policies. This allows an IT organization to optimize the storage infrastructure for maximum return on investment. An ILM strategy should include the following characteristics:

- **Business-centric:** It should be integrated with key processes, applications, and initiatives of the business to meet both current and future growth in information.
- **Centrally managed:** All the information assets of a business should be under the purview of the ILM strategy.
- **Policy-based:** The implementation of ILM should not be restricted to a few departments. ILM should be implemented as a policy and encompass all business applications, processes, and resources.
- **Heterogeneous:** An ILM strategy should take into account all types of storage platforms and operating systems.
- **Optimized:** Because the value of information varies, an ILM strategy should consider the different storage requirements and allocate storage resources based on the information's value to the business.
- **Performance:** All the core elements of the data center should be able to provide optimal performance and service all processing requests at high speed. The infrastructure should be able to support performance requirements.
- **Data integrity:** Data integrity refers to mechanisms such as error correction codes or parity bits which ensure that data is written to disk exactly as it was received. Any variation in data during its retrieval implies corruption, which may affect the operations of the organization.

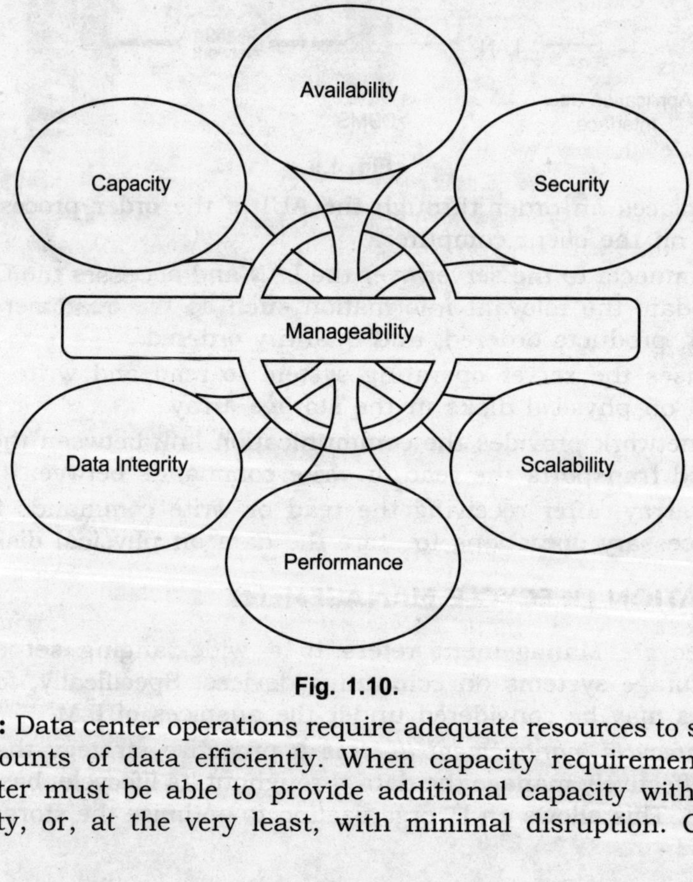

Fig. 1.10.

- **Capacity:** Data center operations require adequate resources to store and process large amounts of data efficiently. When capacity requirements increase, the data center must be able to provide additional capacity without interrupting availability, or, at the very least, with minimal disruption. Capacity may be managed by reallocation of existing resources, rather than by adding new resources.

- **Manageability:** A data center should perform all operations and activities in the most efficient manner. Manageability can be achieved through automation and the reduction of human (manual) intervention in common tasks.

1.6. DATA CATEGORIZATION

Data categorization may refer to:

- Data classification (data management)
- Data classification (business intelligence)
- Classification (machine learning), classification of data using machine learning algorithms
- Assigning a level of sensitivity to classified information
- In computer science, the data type of a piece of data.

In business intelligence, data classification has close ties to data clustering, but where data clustering is descriptive, data classification is predictive. In essence data classification consists of using variables with known values to predict the unknown or future values of other variables. It can be used in *e.g.*, direct marketing, insurance fraud detection or medical diagnosis.

The first step in doing a data classification is to cluster the data set used for category training, to create the wanted number of categories. An algorithm, called the classifier, is then used on the categories, creating a descriptive model for each. These models can then be used to categorize new items in the created classification system.

There are various criteria factors for data categorization:

- **Availability:** All data center elements should be designed to ensure accessibility. The inability of users to access data can have a significant negative impact on a business.
- **Security:** Polices, procedures, and proper integration of the data center core elements that will prevent unauthorized access to information must be established. In addition to the security measures for client access, specific mechanisms must enable servers to access only their allocated resources on storage arrays.
- **Scalability:** Data center operations should be able to allocate additional processing capabilities or storage on demand, without interrupting business operations. Business growth often requires deploying more servers, new applications, and additional databases. The storage solution should be able to grow with the business.

Challenges for Data Classification

There are several challenges in working with data classification. One in particular is that it is necessary for all using categories on, e.g. customers or clients, to do the modeling in an iterative process. This is to make sure that change in the characteristics of customer groups does not go unnoticed, making the existing categories outdated and obsolete, without anyone noticing.

This could be of special importance to insurance or banking companies, where fraud detection is extremely relevant. New fraud patterns may come unnoticed, if the methods to surveil these changes and alert when categories are changing, disappearing or new ones emerge, are not developed and implemented.

Statistical Classification

In machine learning, statistical classification is the problem of identifying the sub-population to which new observations belong, where the identity of the sub-population is unknown, on the basis of a training set of data containing observations whose sub-population is known. Therefore these classifications will show a variable behavior which can be studied by statistics.

Thus, the requirement is that new individual items are placed into groups based on quantitative information on one or more measurements, traits or characteristics, etc. and based on the training set in which previously decided groupings are already established.

The problem here may be contrasted with that for cluster analysis, where the problem is to analyze a single data-set and decide how and whether the observations in the data-set can be divided into groups. In certain terminology, particularly that of machine learning, the classification problem is known as supervised learning, while clustering is known as unsupervised learning.

Unfortunately, terminology can be different in various fields of application. For example, in community ecology, the term "classification" is synonymous with cluster analysis.

Learning Classifiers. Problem Statement

A learning classifier is able to learn based on a sample. The data-set used for training consists of information x and y for each data-point, where x denotes what is generally *a* vector of observed characteristics for the data-item and y denotes a group-label. The label y can take only a finite number of values.

The classification problem can be stated as follows: given training data $\{(x_1, y_1), ..., (x_n, y_n)\}$ produce a rule (or "classifier") h, such that $h(x)$ can be evaluated for any possible value of x (not just those included in the training data) and such that the group attributed to any new observation, specifically

$$\hat{y} = h(x)$$

is as close as possible to the true group label y. For the training data-set, the true labels y_i are known but will not necessarily match their in-sample approximations

$$\hat{y}_i = h(x_i)$$

For new observations, the true labels y_j are unknown, but it is a prime target for the classification procedure that the approximation

$$\hat{y}_j = h(x_j) \approx y_j$$

as well as possible, where the quality of this approximation needs to be judged on the basis of the statistical or probabilistic properties of the overall population from which future observations will be drawn.

SUMMARY

1. Magnetic Tape backup systems, which are still in use in corporate IT centers, but are slow, and aren't good for random access. Random access refers to the ability to

(effectively) access any piece of data by its address (e.g. block number on a hard disk) instantly.

2. Magnetic or hard disks, which are discussed at length below, are a ubiquitous form of high-volume, high-speed, random-access storage.
3. A solid-state drive (SSD) is a data storage device that uses solid-state memory to store persistent data. Unlike flash-based memory cards, an SSD emulates a hard disk drive, thus easily replacing it in most applications. An SSD using SRAM or DRAM (instead of flash memory) is often called a RAM drive. The advantage over (magnetic) disk drives is speed, but the cost per gigabyte is 4 to 5 times that of disk drives, and at the moment the amount of storage is much less per unit.
4. The drive will need power which, for an internal drive, is usual *y* supplied by the computer's own power supply. The disk drive is covered and hermetically sealed, since heads are designed to float above the disk platter with less than a micron of space.
5. The controller will usually plug straight into a slot on the computer motherboard and draw its power from there. This will also let the CPU talk to the host adapter and disks through the system "bus". Sometimes you may find the controlleractually integrated into the computer's motherboard.
6. Data proliferation refers to the prodigious amount of data, structured and unstructured, that businesses and governments continue to generate at an unprecedented rate and the usability problems that result from attempting to store and manage that data. While originally pertaining to problems associated with paper documentation, data proliferation has become a major problem in primary and secondary data storage on computers.
7. Peer-to-Peer Remote Copy) involves two mirror copies of data, one at a primary (local) site and one at a secondary (recovery) site. We say this process is synchronous when data must be successfully written at the secondary system before the write issued by the primary system is acknowledged as complete.
8. Originally, the storage system was just the HDD, but over time storage systems have developed to include advanced technologies that add considerable value to the HDD. Storage systems have evolved to support a variety of added services, as well as connectivity and interface alternatives.
9. The first data storage device was introduced by IBM in 1956. Since then there has been remarkable progress in hard disk drive (HDD) technology, and this has provided the fertile ground on which the entire industry of storage systems has been built.
10. Network-attached storage (NAS) and storage area network (SAN) technologies, and we explain how block virtualization and SAN file systems are necessary to fully reap the benefits of these technologies. We also discuss how the recent trend in storage systems toward managing complexity, ease of-use, and lowering the total cost of ownership has led to the development of autonomic storage.

QUESTION WITH ANSWER

Q.1. Define the types of Storage media.

Ans. There are many types of storage media:

- Flash, which has become a cheap form of fast storage, especially in consumer products.
- Optical storage, which comes in the form of CDs, DVDs and BluRay, etc. These are slow for data access, but still very useful for archives and movies.
- Magnetic Tape backup systems, which are still in use in corporate IT centers, but are slow, and aren't good for random access. Random access refers to the ability to (effectively) access any piece of data by its address (e.g. block number on a hard disk) instantly.
- Magnetic or hard disks, which are discussed at length below, are a ubiquitous form of high-volume, high-speed, random-access storage.

- A solid-state drive (SSD) is a data storage device that uses solid-state memory to store persistent data. Unlike flash-based memory cards, an SSD emulates a hard disk drive, thus easily replacing it in most applications. An SSD using SRAM or DRAM (instead of flash memory) is often called a RAM drive. The advantage over (magnetic) disk drives is speed, but the cost per gigabyte is 4 to 5 times that of disk drives, and at the moment the amount of storage is much less per unit.

Q.2. Define Types of Storage Technology.

Ans. These types of storage can either be static or removable thanks to the ubiquity of USB and firewire. In this paper we will be exclusively talking about magnetic disk (as in hard-drive) storage.

Hard Disk

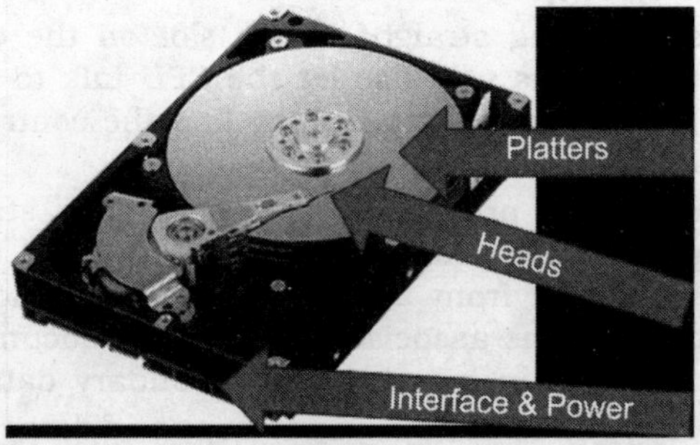

Fig. 1.11.

Hard disks are composed of:

- Multiple spinning magnetic platters that contain the magnetically encoded data. The platters spin around an axle called a "spindle", and sometimes a single drive (like the one shown) can be referred to as such.
- Read/write heads that float above the surface of the platters, usually 2 (top and bottom) for each platter. The heads all move in unison.
- A controler board that drives the heads and can convert I/O requests (commands) into head movement and read/write operations.
- An interface that will be joined to a host adapter board, which for an internal (to a computer) drive, is a separate unit.
- The drive will need power which, for an internal drive, is usually supplied by the computer's own power supply.

Q.3. What is the Data proliferation?

Ans. Data proliferation refers to the prodigious amount of data, structured and unstructured, that businesses and governments continue to generate at an unprecedented rate and the usability problems that result from attempting to store and manage that data. While originally pertaining to problems associated with paper documentation, data proliferation has become a major problem in primary and secondary data storage on computers.

While digital storage has become cheaper, the associated costs, from raw power to maintenance and from metadata to search engines, have not kept up with the proliferation of data. Although the power required to maintain a unit of data has fallen, the cost of facilities which house the digital storage has tended to rise.

The problem of data proliferation is affecting all areas of commerce as the result of the availability of relatively inexpensive data storage devices. This has made it very easy to dump data into secondary storage immediately after its window of usability has passed. This masks problems that could gravely affect the profitability of businesses and the efficient functioning of health services, police and security forces, local and national governments, and many other types of organization.

Q.4. Give overview on storage infrastructure components.

Ans. Storage systems come of age from components to systems. It has long been recognized that the disk drive alone cannot provide the range of storage capabilities required by enterprise systems. The first storage devices were directly controlled by the CPU. Although some advances did take place in the interim, System/360* in 1964 was the first offering of advanced functions in an external storage control unit. The key advantage of a control unit (or controller) was that the I/O commands from the CPU (sometimes called the host) were independently translated into the specific commands necessary to operate the HDD (sometimes called the direct access storage device, or DASD), and so the HDD device itself could be managed independently and asynchronously from the CPU. The control unit included buffering that allowed. The CPU and HDD operations to overlap. Over time there was a greater emphasis on availability, and controllers began to support multiple data paths from CPUs to storage controllers.

Q.5. What do you mean by Disaster recovery.

Ans. Disaster recovery became a requirement for all IT systems, and impacted the design of storage systems as well. The required degree of protection of data requires a solution that may range in technical sophistication from occasional copies onto magnetic tape (and manually transported), through electronic versions of essentially the same principle, to true mirroring solutions and truly distributed systems. Several commonly employed techniques have emerged. A point-in-time copy (offered by IBM under the name lashCopy) is the making of a consistent virtual copy of data as it appeared at a single point in time.

This copy is then kept up to date by following pointers as changes are made. If desired, this virtual copy can, over time, be made into a real copy through physical copying. A second technique, mirroring or continuous copy (offered by IBM under the name.

Peer-to-Peer Remote Copy) involves two mirror copies of data, one at a primary (local) site and one at a secondary (recovery) site. We say this process is synchronous when data must be successfully written at the secondary system before the write issued by the primary system is acknowledged as complete. Although synchronous operation is desirable, it is practical only over limited distances (say, of the order of 100 km). Therefore, other asynchronous schemes, and other significant optimizations, are used to improve the performance of these basic schemes.

Q.6. Explain the Data categorization

Ans. Data categorization may refer to:

- Data classification (data management)
- Data classification (business intelligence)
- Classification (machine learning), classification of data using machine learning algorithms
- Assigning a level of sensitivity to classified information
- In computer science, the data type of a piece of data.

In business intelligence, **data classification** has close ties to data clustering, but where data clustering is descriptive, data classification is predictive. In essence data classification consists of using variables with known values to predict the unknown or future values of other variables. It can be used in e.g. direct marketing, insurance fraud detection or medical diagnosis.

The first step in doing a data classification is to cluster the data set used for category training, to create the wanted number of categories. An algorithm, called the classifier, is then used on the categories, creating a descriptive model for each.

OBJECTIVE TYPE QUESTIONS

1. In a FC SAN, what is an advantage of using copper media instead of optical media?

   (*a*) Better signal-to-noise ratio for short distances

   (*b*) Higher speed

   (*c*) Larger frame payload (*d*) Longer distances are supported

2. A pair of ISLs is configured between two Fibre Channel switches with trunking enabled. The trunk is carrying the I/O traffic from multiple hosts generating a total of 3,200 IOPS each, with a block size of 4 KB.
   Assuming 2 Gb links, which percentage represents the best estimate of ISL utilization?
   (*a*) 1 (*b*) 3
   (*c*) 10 (*d*) 20
3. A pair of ISLs is configured between two Fibre Channel switches with trunking enabled. The trunk is carrying the I/O traffic from several hosts generating a total of 3,000 IOPS with a block size of 32 KB.
   Assuming 2 Gb links, what is the ISL utilization?
   (*a*) 3% (*b*) 25%
   (*c*) 50% (*d*) 90%
4. In a requirements analysis, what are classified as external requirements?
   (*a*) Availability and reliability (*b*) Delivery and standards
   (*c*) Portability and interoperability (*d*) Privacy and legislative
5. What is a benefit of Kerberos over other authentication protocols?
   (*a*) Client can provide its own privilege information, reducing authentication traffic
   (*b*) Ensures a user will not be disconnected from a share because tickets do not expire
   (*c*) Stores file system ACLs for the domain environment
   (*d*) User ID and password information is never stored on the Kerberos server

ANSWERS

1. (a) 2. (b) 3. (b) 4. (d) 5. (a)

CHAPTER

2

STORAGE SYSTEMS ARCHITECTURE

2.0. INTRODUCTION

With network content expected to continue expanding at explosive rates over the next five years, analysts predict that enterprise storage will account for 75% of all computer hardware expenditures. Such expansion makes it increasingly important for IT professionals to develop comprehensive strategies designed to optimize network infrastructure with storage solutions that will enable scalability, reliability, performance, availability, affordability and manageability.

Compounding the challenge of explosive content demands, there are two major technology shifts IT professionals must consider when developing an enterprise storage strategy:

- first, the impact of networking technology on storage architecture and content management
- second, the impact of parallel processing on the design of storage products.

These two technology shifts have produced three mutually coexistent methods for connecting storage to computing platforms: Direct Attached Storage (DAS), Network Attached Storage (NAS), and Storage Area Networks (SAN).

Inside this Chapter

- Introduction
- Storage Systems Architecture
- Intelligent Disk system overview
- Contrast of integrated vs. modular arrays
- Component architecture of Intelligent disk subsystem
- Disk Physical Structure components
- Properties of physical structure components
- Performance and specifications

2.1. STORAGE SYSTEMS ARCHITECTURE

Computer data storage, often called storage or memory, refers to computer components and recording media that retain digital data. Data storage is one of the core functions and fundamental components of computers.

In contemporary usage, memory usually refers to semiconductor storage random-access memory, typically DRAM (Dynamic-RAM). Memory can refer to other forms of fast but temporary storage. Storage refers to storage devices and their media not directly accessible by the CPU, (secondary or tertiary storage)—typically hard disk drives, optical disc drives, and other devices slower than RAM but more permanent. Historically, memory has been called *main memory, real storage or internal memory* while storage devices have been referred to as *secondary storage, external memory or auxiliary/peripheral storage.*

The *physical components* are the hardware elements that connect the host to storage and the *logical components* of connectivity are the protocols used for communication between the host and storage. The three physical components of connectivity between the host and storage are Bus, Port, and Cable. The *bus* is the collection of paths that facilitates data transmission from one part of a computer to another, such as from the CPU to the memory. The *port* is a specialized outlet that enables connectivity between the host and external devices. *Cables* connect hosts to internal or external devices using copper or fiber optic media.

Physical components communicate across a bus by sending bits (control, data, and address) of data between devices. These bits are transmitted through the bus in either of the following ways:

- **Serially:** Bits are transmitted sequentially along a single path. This transmission can be unidirectional or bidirectional.
- **In parallel:** Bits are transmitted along multiple paths simultaneously. Parallel can also be bidirectional.

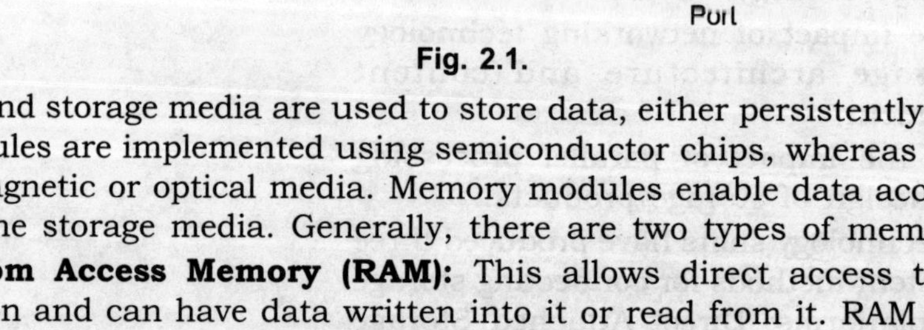

Fig. 2.1.

Memory and storage media are used to store data, either persistently or temporarily. Memory modules are implemented using semiconductor chips, whereas storage devices use either magnetic or optical media. Memory modules enable data access at a higher speed than the storage media. Generally, there are two types of memory on a host:

- **Random Access Memory (RAM):** This allows direct access to any memory location and can have data written into it or read from it. RAM is volatile; this type of memory requires a constant supply of power to maintain memory cell content. Data is erased when the system's power is turned off or interrupted.
- **Read-Only Memory (ROM):** Non-volatile and only allows data to be read from it. ROM holds data for execution of internal routines, such as system startup.

Storage devices are less expensive than semiconductor memory. Examples of storage devices are as follows:

- Hard disk (magnetic)
- CD-ROM or DVD-ROM (optical)
- Floppy disk (magnetic)
- Tape drive (magnetic)

Generally, the lower a storage is in the hierarchy, the lesser its bandwidth and the greater its access latency is from the CPU. This traditional division of storage to primary, secondary, tertiary and off-line storage is also guided by cost per bit.

Hierarchy of Storage

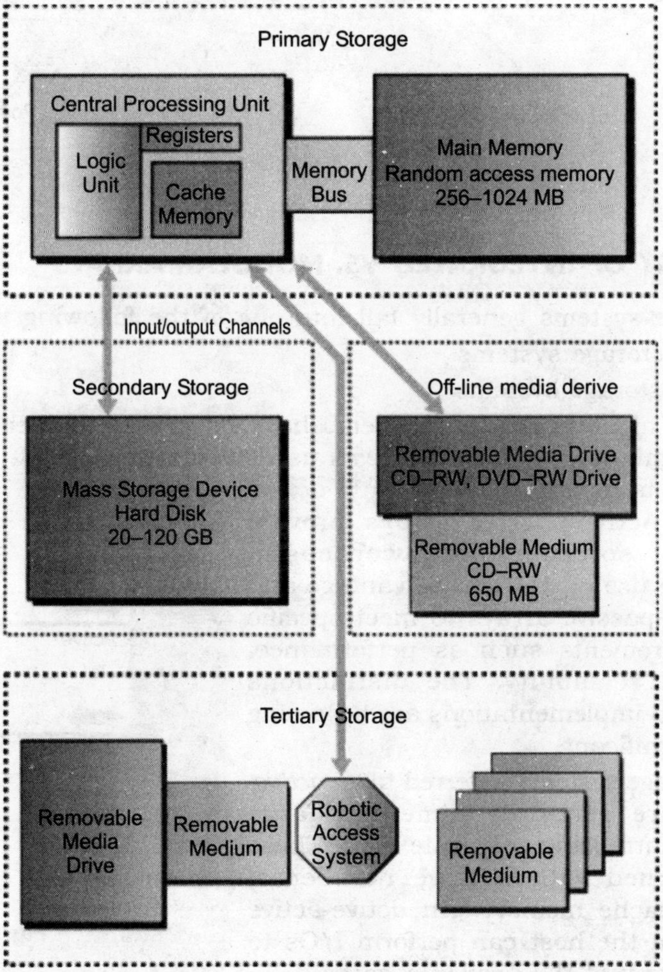

Fig. 2.2.

2.2. INTELLIGENT DISK SYSTEM OVERVIEW

An intelligent storage system consists of four key components: *find end*, *cache*, *back end*, and *physical disks*. Figure 2.3, illustrates these components and their interconnections. An I/O request received from the host at the front-end port is processed through cache and the back end, to enable storage and retrieval of data from the physical disk. A read request can be serviced directly from cache if the requested data is found in cache.

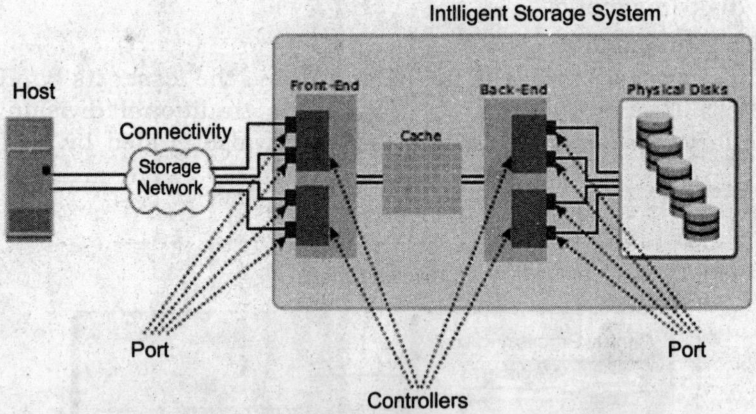

Fig. 2.3.

2.3. CONTRAST OF INTEGRATED VS. MODULAR ARRAYS

Intelligent storage systems generally fall into one of the following two categories:

- High-end storage systems
- Midrange storage systems

Traditionally, high-end storage systems have been implemented with *active-active arrays,* whereas midrange *storage systems* used typically in small- and mediumsized enterprises have been implemented with *active-passive arrays.* Active-passive arrays provide optimal storage solutions at lower costs. Enterprises make use of this cost advantage and implement active-passive arrays to meet specific application requirements such as performance, availability, and scalability. The distinctions between these two implementations are becoming increasingly insignificant.

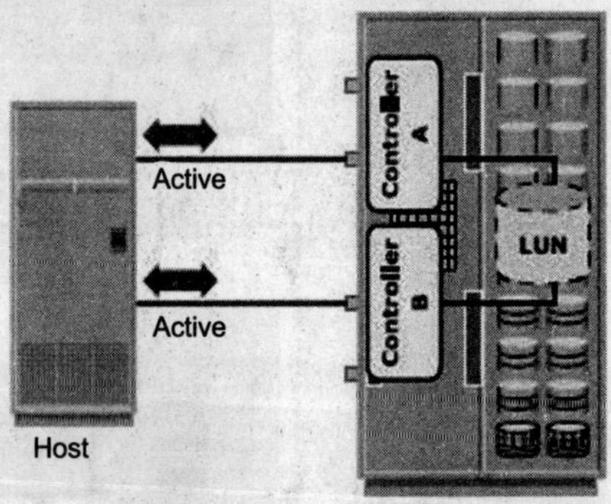

Fig. 2.4.

High-end storage systems, referred to as *active-active arrays,* are generally aimed at large enterprises for centralizing corporate data. These arrays are designed with a large number of controllers and cache memory. An active-active array implies that the host can perform I/Os to its LUNs across any of the available paths.

To address the enterprise storage needs, these arrays provide the following capabilities:

- Large storage capacity
- Large amounts of cache to service host I/Os optimally
- Fault tolerance architecture to improve data availability

- Connectivity to mainframe computers and open systems hosts.
- Availability of multiple front-end ports and interface protocols to serve a large number of hosts.
- Availability of multiple back-end Fibre Channel or SCSI RAID controllers to manage disk processing.
- Scalability to support increased connectivity, performance, and storage capacity requirements.
- Ability to handle large amounts of concurrent I/Os from a number of servers and applications.
- Support for array-based local and remote replication.

In addition to these features, high-end arrays possess some unique features and functionals that are required for mission-critical applications in large enterprises.

Midrange storage systems are also referred to as *active-passive arrays* and they are best suited for small- and medium-sized enterprises. In an active-passive array, a host can perform I/Os to a LUN only through the paths to the owning controller of that LUN. These paths are called *active paths.* The other paths are passive with respect to this LUN. As shown in Fig. 2.5, the host can perform reads or writes to the LUN only through the path to controller A, as controller A is the owner of that LUN. The path to controller B remains passive and no I/O activity is performed through this path.

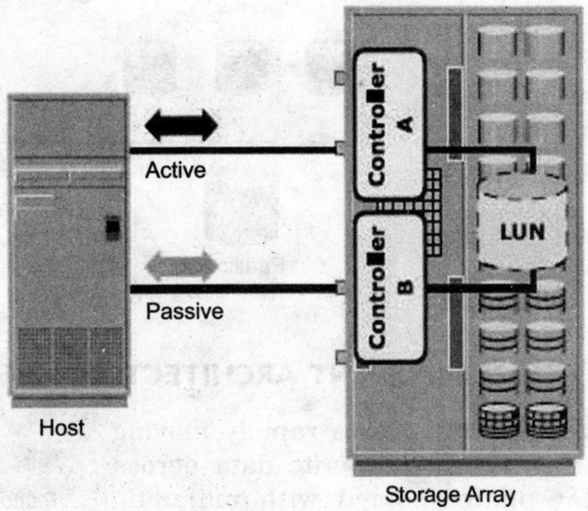

Fig. 2.5.

Midrange storage systems are typically designed with two controllers, each of which contains host interfaces, cache, RAID controllers, and disk drive interfaces.

Midrange arrays are designed to meet the requirements of small and medium enterprises; therefore, they host less storage capacity and global cache than active-active arrays. There are also fewer front-end ports for connection to servers. However, they ensure high redundancy and high performance for applications with predictable workloads. They also support array-based local and remote replication.

The evolution of *Logical Volume Managers* (*LVMs*) enabled the dynamic extension of file system, capacity and efficient storage management. LVM is software that runs on the host computer and manages the logical and physical storage. LVM is an optional, intermediate layer between the file system and the physical disk. It can aggregate several smaller disks to form a larger virtual disk or to partition a larger-capacity disk into virtual, smaller-capacity disks, which are then presented to applications. The LVM provides optimized storage access and simplifies storage resource management. It hides details about the physical disk and location of data on the disk, and it enables administrators to change the storage allocation without changing the hardware, even when the application is running.

The basic LVM components are the *physical volumes*, *volume groups*, and *logical volumes*. In LVM terminology, each physical disk connected to the host sytem is a *physical volume* (*PV*). LVM converts the physical storage provided by the physical volumes to a logical view of storage, which is then used by the operating system and applications. A *volume group* is created by grouping together one or more physical volumes. A unique *physical volume identifier* (*PVID*) is assigned to each physical volume when it is initialized for use by the LVM. Physical volumes can be added or removed from a volume group dynamically. They cannot be shared between volume groups; the entire physical volume becomes part of a volume group. Each physical volume is partitioned into equal-sized data blocks called *physical extents* when the volume group is created.

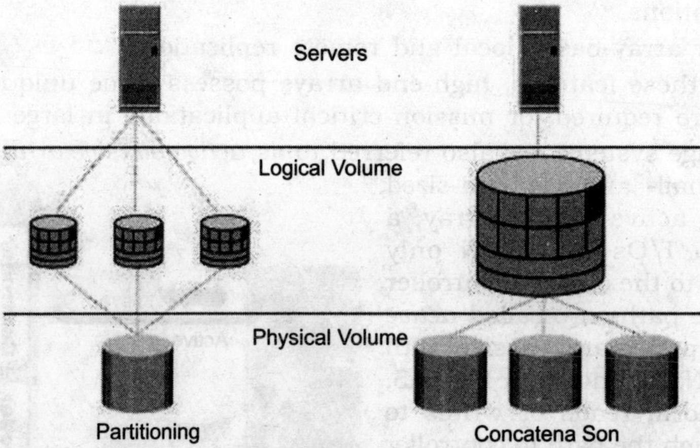

Fig. 2.6.

2.4. COMPONENT ARCHITECTURE OF INTELLIGENT DISK SUBSYSTEM

A disk drive uses a rapidly moving arm to read and write data across flat platter coated with magnetic particlaes. Data is transferred from the magnetic platter through the R/W head to the computer. Several platters are assembled together with the R/W head and controller, most commonly referred to as a *hard disk drive* (*HDD*). Data can be recorded and erased on a magnetic disk any number of times. This section details the different components of the disk, the mechanism for organizing and storing data on disks, and the factors that affect disk performance.

Key components of a disk drive are *platter*, *spindle*, *read/write head*, *actuator arm assembly*, and *controller* (Fig. 2.7.)

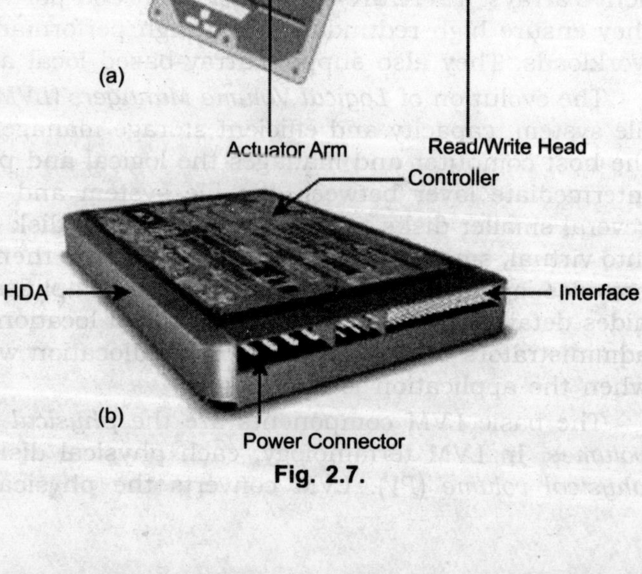

Fig. 2.7.

2.5. DISK PHYSICAL STRUCTURE COMPONENTS

2.5.1. Read/Write (R/W) heads

Read/ Write (R/ W) heads, shown in Fig. 2.8, read and write data from or to a platter. Drives have two R/W heads per platter, one for each surface of the platter. The R/W head changes the magnetic polarization on the surface of.

2.5.2. Spindle

A spindle connects all the platters, as shown in Fig. 2.8, and is connected to a motor. The motor of the spindle rotates with a constant speed.

The disk platter spins at a speed of several thousands of revolutions per minute (rpm). Disk drives have spindle speeds of 7,200 rpm, 10,000 rpm, or 15,000 rpm. Disk used on current storage systems have a platter diameter of 3.5″ (90 mm). When the platter spins at 15,000 rpm, the outer edge is moving at around 25 percent of the speed of sound. The speed of the platter is increaseing with improvements in technology, although the extent to which it can be improved is limited.

2.5.3. Platter

The platter when writing data. While reading data, this head detects magnetic polarization on the surface of the platter. During reads and writes, the R/W head senses the magnetic polarization and never touches the surface of the platter. When the spindle is rotating, there is a microscopic air gap between the R/W heads and the platters, known as the *head flying height*. This air gap is removed when the spindle stops rotating and the R/W head rests on a special area on the platter near the spindle. This area is called the *landing zone*. The landing zone is coated with a lubricant to reduce friction between the head and the platter. The logic on the disk drive ensures that heads are moved to the landing zone before they touch the surface. If the drive malfunctions and the R/W head accidentally touches the surface of the platter outside the landing zone, a *head crash* occurs. In a head crash, the magnetic coating on the platter is scratched and may cause damage to the R/W head. A head crash generally results in data loss.

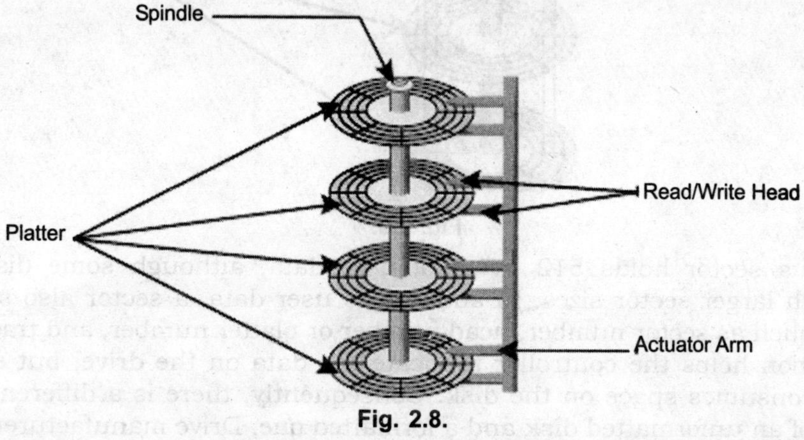

Fig. 2.8.

The R/W heads are mounted on the *actuator arm assembly* which positions the R/W head at the location on the platter where the data needs to be written or read. The R/W heads for all platters on a drive are attached to one actuator arm assembly and move across the platters simultaneously. Note that there are two R/W heads per platter, one for each surface,

2.5.4. Controller

The *controller* is a printed circuit board, mounted at the bottom of a disk drive. It consists of a microprocessor, internal memory, circuitry, and firmware. The firmware controls power to the spindle motor and the speed of the motor. It also manages communication between the drive and the host. In addition, it controls the R/W operations by moving the actuator arm and switching between different R/W heads, and performs the optimization of data access.

Data on the disk is recorded on *tracks*, which are concentric rings on the platter around the spindle, as shown in Fig. 2.9. The tracks are numbered, starting from zero, from the outer edge of the platter. The number of *tracks per inch (TPI)* on the platter (or the *track density*) measures how tightly the tracks are packed on a platter.

Each track is divided into smaller units called *sectors*. A sector is the smallest, individually addressable unit of storage. The track and sector structure is written on the platter by the drive manufacturer using a formatting operation. The number of sectors per track varies according to the specific drive. The first personal computer disks had 17 sectors per track. Recent disks have a much larger number of sectors on a single track. There can be thousands of tracks on a platter, depending on the physical dimensions and recording density of the platter.

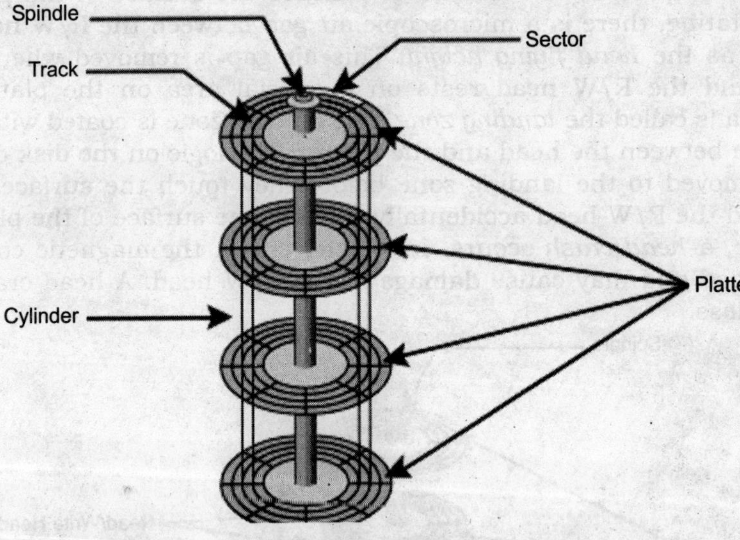

Fig. 2.9.

Typically, a sector holds 512 bytes of user data, although some disks can be formatted with larger sector sizes. In addition to user data, a sector also stores other information, such as sector number, head number or platter number, and track number. This information helps the controller to locate the data on the drive, but storing this information consumes space on the disk. Consequently, there is a difference between the capacity of an unformatted disk and a formatted one. Drive manufacturers generally advertise the unformatted capacity — for example, a disk advertised as being 500 GB will only hold 465.7 GB of user data, and the remaining 34.3 GB is used for *metadata*.

2.5.5. Cylinder

A cylinder is the set of identical tracks on both surfaces of each drive platter. The location of drive heads is referred to by cylinder number, not by track number.

2.6. PROPERTIES OF PHYSICAL STRUCTURE COMPONENTS

A disk drive is an electromechanical device that governs the overall performance of the storage system environment. The various factors that affect the performance of disk drives are discussed in this section.

2.6.1. Disk Service Time

Disk service time is the time taken by a disk to complete an I/O request. Components that contribute to service time on a disk drive are *seek time, rotational latency,* and *data transfer rate.*

2.6.1.1. Seek Time

The *seek time* (also called *access time*) describes the time taken to position the R/W heads across the platter with a radial movement (moving along the radius of the platter). In other words, it is the time taken to reposition and settle the arm and the head over the correct track. The lower the seek time, the faster the I/O operation.

Disk vendors publish the following seek time specifications:

- **Full Stroke:** The time taken by the R/W head to move across the entire width of the disk, from the innermost track to the outermost track.
- **Average:** The average time taken by the R/W head to move from one random track to another, normally listed as the time for one-third of a full stroke.
- **Track-to-Track:** The time taken by the R/W head to move between adjacent tracks.

Each of these specifications is measured in milliseconds. The average seek time on a modern disk is typically in the range of 3 to 15 milliseconds. Seek time has more impact on the read operation of random tracks rather than adjacent tracks. To minimize the seek time, data can be written to only a subset of the available cylinders. This results in lower usable capacity than the actual capacity of the drive. For example, a 500 GB disk drive is set up to use only the first 40 percent of the cylinders and is effectively treated as a 200 GB drive. This is known as *short-stroking* the drive.

2.6.1.2. Rotational Latency

To access data, the actuator arm moves the R/W head over the platter to a particular track while the platter spins to position the requested sector under the R/W head. The time taken by the platter to rotate and position the data under the R/W head is called *rotational latency.* This latency depends on the rotation speed of the spindle and is measured in milliseconds. The average rotational latency is one-half of the time taken for a full rotation. Similar to the seek time, rotational latency has more impact on the reading/writing of random sectors on the disk than on the same operations on adjacent sectors.

Average rotational latency is around 5.5 ms for a 5,400-rpm drive, and around 2.0 ms for a 15,000-rpm drive.

2.6.1.3. Data Transfer Rate

The *data transfer rate* (also called *transfer rate*) refers to the average amount of data per unit time that the drive can deliver to the HBA. It is important to first understand the process of read and write operations in order to calculate data transfer rates. In a *read operation*, the data first moves from disk platters to R/W heads, and then it moves to the drive's internal *buffer*. Finally, data moves from the buffer through the interface to the host HBA. In a *write operation*, the data moves from the HBA to the internal buffer of the disk drive through the drive's interface. The data then moves from the buffer to the R/W heads. Finally, it moves from the R/W heads to the platters.

The data transfer rates during the R/W operations are measured in terms of internal and external transfer rates, as shown in Fig. 2.10.

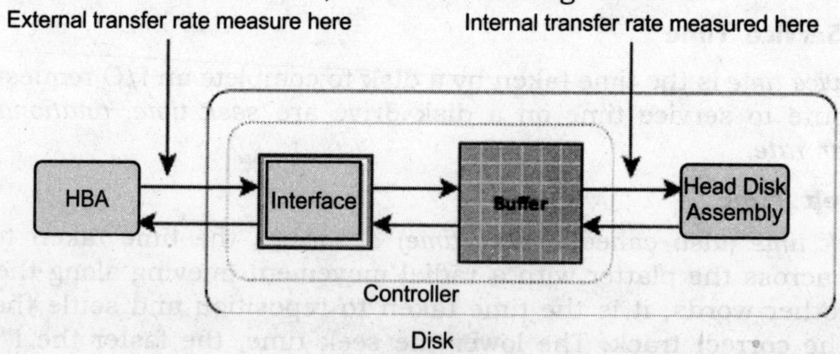

Fig. 2.10.

Internal transfer rate is the speed at which data moves from a single track of a platter's surface to internal buffer (cache) of the disk. Internal transfer rate takes into account factors such as the seek time. *External transfer rate* is the rate at which data can be moved through the interface to the HBA. External transfer rate is generally the advertised speed of the interface, such as 133 MB/s for ATA. The sustained external transfer rate is lower than the interface speed.

2.7. PERFORMANCE AND SPECIFICATIONS

To understand the laws of disk performance, a disk can be viewed as a black box consisting of two elements:

- **Queue:** The location where an I/O request waits before it is processed by the I/O controller.
- **Disk I/O Controller:** Processes I/Os that are waiting in the queue one by one.

The I/O requests arrive at the controller at the rate generated by the application. This rate is also called the *arrival rate*. These requests are held in the I/O queue, and the I/O controller processes them one by one, as shown in Fig. 2.11.

The I/O arrival rate, the queue length, and the time taken by the I/O controller to process each request determines the performance of the disk system, which is measured in terms of response time.

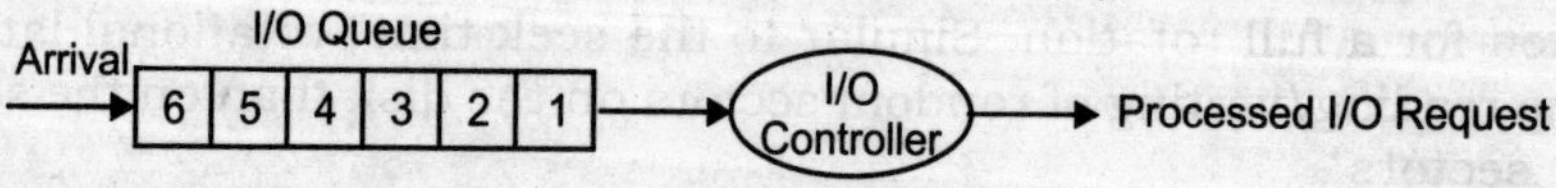

Fig. 2.11.

Little's Law is a fundamental law describing the relationship between the number of requests in a queue and the response time. The law states the following relation (numbers in parentheses indicate the equation number for cross-referencing):

$$N = a \times R \qquad ...(1)$$

where

"N" is the total number of request in the quening system (requests in the queue + requests in the I/O controller)

"a" is the arrival rate, or the number of I/O requests that arrive to the system per unit of time.

"R" is the average response time or the turnaround time for an I/O request—the total time from arrival to departure from the system.

The *utilization law* is another important law that defined the I/O controller utilizationl. This law states the relation:

$$U = a \times R_S \qquad ...(2)$$

where

"U" is the I/O controller utilization

"R_S" is the *service time*, or the average time spent by a request on the controller, $1/R_S$ is the *service rate.*

From the arrival rate "a", the average inter-arrival time, R_S can be computed as:

$$R_a = 1/a \qquad ...(3)$$

Consequently, *utilization* can be defined as the ratio of the service time to the average inter-arrival time, and is expressed as:

$$U = R_S/R_a \qquad ...(4)$$

The value of this ratio varies between 0 and 1.

Here, it is important to realize that in a single controller system, the arrival rate must be smaller than the service rate. In other words, the service time must be smaller than the average inter-arrival time; otherwise, I/O requests will arrive into the system faster than the I/O controller can process them.

With the help of these two fundamental laws, a number of important measures of disk performance, such as average response time, average queue length, and time spent by a request in a queue can be derived.

In the following equation, the term average response rate (S) can be defined as the reciprocal of the average response time (R), and is derived as follows:

$$S = \text{service rate} - \text{arrival rate}$$

Consequently,

$$R = 1/(\text{service rate} - \text{arrival rate})$$

$$R = 1/(R_S - 1/R_a)$$

$$= 1/(1/R_S - a) \qquad \text{(from eq. 3)}$$

$$= R_S/(1 - a \times R_S)$$

$$R = R_S/(1 - U) \qquad ...(5) \quad \text{(from eq. 2)}$$

As a result,

Average response time (R) = service time/(1 – utilization) (from equation 2)

As utilization reaches 1 – that is, as the I/O controller saturates—the response time is closer to infinity. In essence, the saturated component, or the bottleneck, forces the serialization of I/O requests, meaning each I/O request has to wait for the completion of the I/O requests that preceded it.

Utilization (U) can also be used to represent the average number of I/O requests on the controller, as shown in the following:

Number of requests in the queue (N_Q) = Number of requests in the system (N)

Number of requests on the controller or utilization (U). Number of requests in a queue is also referred to as *average queue size.*

$$N_Q = N - U$$
$$= a \times R - U \qquad \text{(from eq. 1)}$$
$$= a \times (R_S / (1 - U)) - U \qquad \text{(from eq. 5)}$$
$$= (R_S/R_a) / (1 - U) - U \qquad \text{(from eq. 3)}$$
$$= U / (1 - U) - U \qquad \text{(from eq. 4)}$$
$$= U (1 / (1 - U) - 1)$$
$$= U^2 / (1 - U) \qquad ...(6)$$

The time spent by a request in the queue is equal to the time spent by a request in the system, or the average response time minutes the time spent by a request on the controller for processing:

$$= R_S / (1 - U) - R_S \qquad \text{(from eq. 5)}$$
$$= U \times R_S / (1 - U)$$
$$= U \times \text{average response time}$$
$$= \text{Utilization} \times R \qquad ...(7)$$

Consider a disk I/O system in which an I/O request arrives at a rate of 100 I/Os per second. The service time, R_S, is 8 ms. The following measures of disk performance can be computed using the relationships developed above – utilization of I/O controller (U), total response time (R), average queue size [U^2 / (1 – U)] and total time spent by a request in a queue ($U \times R$), as follows:

Arrival rate (a) = 100/I/O/s; consequently, the arrival time

$$R_a = 1/a = 10 \text{ ms}$$
$$R_S = 8 \text{ ms} \quad \text{(given)}$$

1. Utilization (U) = R_s/R_a = 8/10 = 0.8 or 80%
2. Response time (R) = $R_S/(1 - U)$ = 8/(1 – 0.8) = 40 ms
3. Average queue size = $U^2/(1 - U)$ = (0.8)2/(1 – 0.8) = 3.2
4. Time spent by a request in a queue = $U \times R$, or the total response time-service time = 32 ms

Now, if controller power is doubled, the service time is halved; consequently, R_S = 4 ms in this scenario.

1. Utilization (U) = 4/10 = 0.4 or 40%
2. Response time (R) = 4/(1 – 0.4) = 6.67 ms
3. Average queue size = $(0.4)^2/(1 - 0.4)$ = 0.26
4. Time spent by a request in a queue = 0.4 × 6.67 = 2.67 ms

SUMMARY

1. With network content expected to continue expanding at explosive rates over the next five years, analysts predict that enterprise storage will account for 75 % of all computer hardware expenditures.
2. Computer data storage, often called storage or memory, refers to computer components and recording media that retain digital data. Data storage is one of the core functions and fundamental components of computers.
3. In contemporary usage, memory usually refers to semiconductor storage random-access memory, typically DRAM (Dynamic-RAM). Memory can refer to other forms of fast but temporary storage. Storage refers to storage devices and their media not directly accessible

by the CPU, (secondary or tertiary storage)—typically hard disk drives, optical disc drives, and other devices slower than RAM but more permanent.

4. A digital computer represents data using the binary numeral system. Text, numbers, pictures, audio, and nearly any other form of information can be converted into a string of bits, or binary digits, each of which has a value of 1 or 0. The most common unit of storage is the byte, equal to 8 bits. A piece of information can be handled by any computer whose storage space is large enough to accommodate the binary representation of the piece of information, or simply data. For example, using eight million bits, or about one megabyte, a typical computer could store a short novel.
5. Primary storage (or main memory or internal memory), often referred to simply as memory, is the only one directly accessible to the CPU. The CPU continuously reads instructions stored there and executes them as required. Any data actively operated on is also stored there in uniform manner.
6. Processor registers are located inside the processor. Each register typically holds a word of data (often 32 or 64 bits). CPU instructions instruct the arithmetic and logic unit to perform various calculations or other operations on this data (or with the help of it). Registers are the fastest of all forms of computer data storage.
7. Many types of "ROM" are not literally read only, as updates are possible; however it is slow and memory must be erased in large portions before it can be re-written. Some embedded systems run programs directly from ROM (or similar), because such programs are rarely changed. Standard computers do not store non-rudimentary programs in ROM, rather use large capacities of secondary storage, which is non-volatile as well, and not as costly.
8. The secondary storage is often formatted according to a file system format, which provides the abstraction necessary to organize data into files and directories, providing also additional information (called metadata) describing the owner of a certain file, the access time, the access permissions, and other information.
9. The time it takes to access a particular location in storage. The relevant unit of measurement is typically nanosecond for primary storage, millisecond for secondary storage, and second for tertiary storage. It may make sense to separate read latency and write latency, and in case of sequential access storage, minimum, maximum and average latency.
10. Optical storage, the typical optical disc, stores information in deformities on the surface of a circular disc and reads this information by illuminating the surface with a laser diode and observing the reflection. Optical disc storage is non-volatile. The deformities may be permanent (read only media), formed once (write once media) or reversible (recordable or read/write media).

QUESTION WITH ANSWER

Q.1. What is Storage Systems Architecture?

Ans. Computer data storage, often called storage or memory, refers to computer components and recording media that retain digital data. Data storage is one of the core functions and fundamental components of computers.

In contemporary usage, memory usually refers to semiconductor storage random-access memory, typically DRAM (Dynamic-RAM). Memory can refer to other forms of fast but temporary storage. Storage refers to storage devices and their media not directly accessible by the CPU, (secondary or tertiary storage)—typically hard disk drives, optical disc drives, and other devices slower than RAM but more permanent. Historically, memory has been called main memory, real storage or internal memory while storage devices have been referred to as secondary storage, external memory or auxiliary/peripheral storage.

The distinctions are fundamental to the architecture of computers. The distinctions also reflect an important and significant technical difference between memory and mass storage devices, which has been blurred by the historical usage of the term storage. Nevertheless, this article uses the traditional nomenclature.

Many different forms of storage, based on various natural phenomena, have been invented. So far, no practical universal storage medium exists, and all forms of storage have some drawbacks. Therefore a computer system usually contains several kinds of storage, each with an individual purpose.

Q.2. Briefly Explain the Hierarchy of storage.

Ans. Various forms of storage, divided according to their distance from the central processing unit. The fundamental components of a general-purpose computer are arithmetic and logic unit, control circuitry, storage space, and input/output devices. Technology and capacity as in common home computers around 2005.

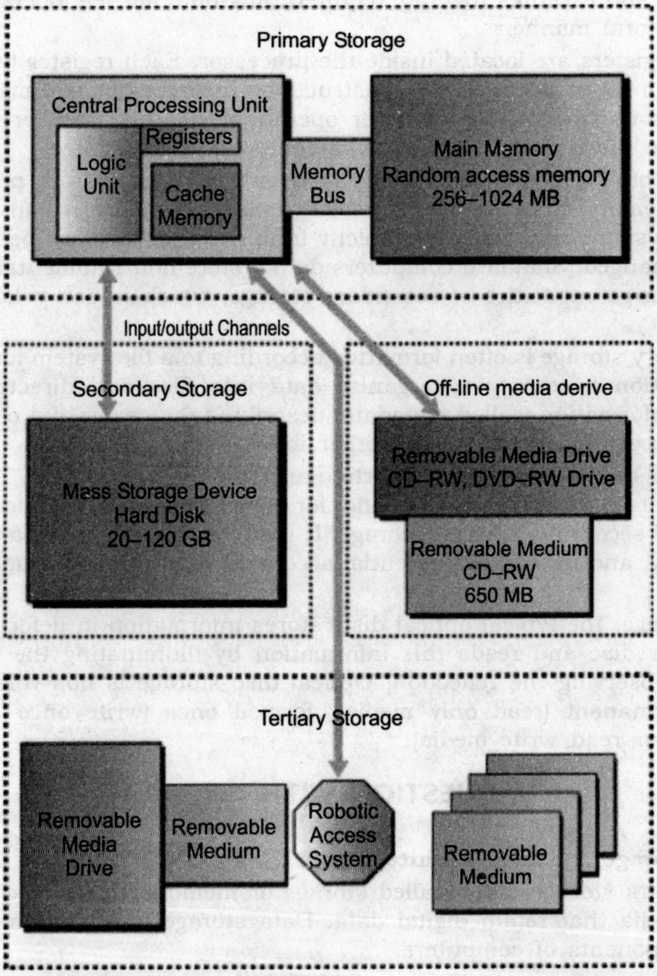

Fig. 2.12.

Q.3. Define Primary storage, Briefly.

Ans. Primary storage (or main memory or internal memory), often referred to simply as memory, is the only one directly accessible to the CPU. The CPU continuously reads instructions stored there and executes them as required. Any data actively operated on is also stored there in uniform manner.

Historically, early computers used delay lines, Williams tubes, or rotating magnetic drums as primary storage. By 1954, those unreliable methods were mostly replaced by magnetic core memory. Core memory remained dominant until the 1970s, when advances in integrated circuit technology allowed semiconductor memory to become economically competitive.

Many types of "ROM" are not literally read only, as updates are possible; however it is slow and memory must be erased in large portions before it can be re-written. Some embedded systems run programs directly from ROM (or similar), because such programs are rarely changed. Standard computers do not store non-rudimentary programs in ROM, rather use large capacities of secondary storage, which is non-volatile as well, and not as costly.

Recently, primary storage and secondary storage in some uses refer to what was historically called, respectively, secondary storage and tertiary storage.

Q.4. Briefly Explain Secondary storage with example.

Ans. Secondary storage (also known as external memory or auxiliary storage), differs from primary storage in that it is not directly accessible by the CPU. The computer usually uses its input/output channels to access secondary storage and transfers the desired data using intermediate area in primary storage. Secondary storage does not lose the data when the device is powered down-it is non-volatile. Per unit, it is typically also two orders of magnitude less expensive than primary storage. Consequently, modern computer systems typically have two orders of magnitude more secondary storage than primary storage and data is kept for a longer time there.

In modern computers, hard disk drives are usually used as secondary storage. The time taken to access a given byte of information stored on a hard disk is typically a few thousandths of a second, or milliseconds. By contrast, the time taken to access a given byte of information stored in random access memory is measured in billionths of a second, or nanoseconds. This illustrates the significant access-time difference which distinguishes solid-state memory from rotating magnetic storage devices: hard disks are typically about a million times slower than memory. Rotating optical storage devices, such as CD and DVD drives, have even longer access times. With disk drives, once the disk read/write head reaches the proper placement and the data of interest rotates under it, subsequent data on the track are very fast to access. As a result, in order to hide the initial seek time and rotational latency, data is transferred to and from disks in large contiguous blocks.

Q.5. Explain Characteristics of storage device.

Ans. Storage technologies at all levels of the storage hierarchy can be differentiated by evaluating certain core characteristics as well as measuring characteristics specific to a particular implementation. These core characteristics are volatility, mutability, accessibility, and addressability. For any particular implementation of any storage technology, the characteristics worth measuring are capacity and performance.

Volatility: Non-volatile memory Will retain the stored information even if it is not constantly supplied with electric power. It is suitable for long-term storage of information.

Volatile memory: Requires constant power to maintain the stored information. The fastest memory technologies of today are volatile ones (not a universal rule). Since primary storage is required to be very fast, it predominantly uses volatile memory.

Dynamic random-access memory: A form of volatile memory which also requires the stored information to be periodically re-read and re-written, or refreshed, otherwise it would vanish.

A form of volatile memory similar to DRAM with the exception that it never needs to be refreshed as long as power is applied. (It loses its content if power is removed).

Mutability: Read/write storage or mutable storage.

Allows information to be overwritten at any time. A computer without some amount of read/write storage for primary storage purposes would be useless for many tasks. Modern computers typically use read/write storage also for secondary storage.

Read only storage: Retains the information stored at the time of manufacture, and write once storage (Write Once Read Many) allows the information to be written only once at some point after manufacture. These are called immutable storage. Immutable storage is used for tertiary and off-line storage. Examples include CD-ROM and CD-R.

Slow write, fast read storage: Read/write storage which allows information to be overwritten multiple times, but with the write operation being much slower than the read operation. Examples include CD-RW and flash memory.

Q.6. Briefly Explain the Performance of storage device.

Ans. Latency: The time it takes to access a particular location in storage. The relevant unit of measurement is typically nanosecond for primary storage, millisecond for secondary storage, and second for tertiary storage. It may make sense to separate read latency and write latency, and in case of sequential access storage, minimum, maximum and average latency.

Throughput: The rate at which information can be read from or written to the storage. In computer data storage, throughput is usually expressed in terms of megabytes per second or MB/s, though bit rate may also be used. As with latency, read rate and write rate may need to be differentiated. Also accessing media sequentially, as opposed to randomly, typically yields maximum throughput.

Energy use:

- Storage devices that reduce fan usage, automatically shut-down during inactivity, and low power hard drives can reduce energy consumption 90 percent.
- 2.5 inch hard disk drives often consume less power than larger ones. Low capacity solid-state drives have no moving parts and consume less power than hard disks. Also, memory may use more power than hard disks.

OBJECTIVE TYPE QUESTIONS

1. Click the Exhibit button.
What is the name of the pictured connector used for Ethernet connections?
(*a*) RJ-8 (*b*) RJ-11
(*c*) RJ-45 (*d*) RJ-48

2. A company has decided to use all of SAN Advisors features. What will this company also need?
(*a*) An Internet connection, Oracle 8, and Excel 2002
(*b*) An Internet connection, Visio 2002, and Excel 2002.
(*c*) Microsoft SQL 2000 and Excel 2002
(*d*) Oracle 8 and Excel 2002

3. In CLARiiON environments, to perform management from a host that has an agent installed, the storage must be added as a Privileged User in a specific file. Which file is this?
(*a*) agent.cfg (*b*) agent.config
(*c*) naviagent.cfg (*d*) naviagent.config

4. What is a valid CLARiiON World Wide Node Name?
(*a*) 50 : 06 : 01 : 63 : 39 : a0 : 0d : 84
(*b*) 50 : 06 : 01 : 60 : b9 : a0 : 0d : 84
(*c*) 50 : 06 : 01 : 6b : 39 : a0 : 0d : 84
(*d*) D.21 : 00 : 00 : e0 : 8b : 86 : 88 : db

5. Which architecture allows for initiators to exist in an Ethernet environment while the target storage exists in a Fibre Channel environment?
(*a*) Bridging (*b*) Routing
(*c*) Switching (*d*) Tunneling

ANSWERS

1. (*b*) 2. (*b*) 3. (*b*) 4. (*b*) 5. (*a*)

CHAPTER

3

RAID LEVELS AND PARITY ALGORITHMS

3.0. INTRODUCTION

Hard disks and tapes are currently the most important media for the storage of data. Whenstorage networks are introduced, the existing small storage devices are replaced by a few large storage systems (storage consolidation). For example, individual hard disks and small disk stacks are replaced by large disk subsystems that can store between a few hundred gigabytes and several ten terabytes of data, depending upon size. Furthermore, they have the advantage that functions such as high availability, high performance, instant copies and remote mirroring are available at a reasonable price even in the ûeld of open systems (UNIX, Windows NT/2000, OS/400, Novell Netware, MacOS). The administration of a few large storage systems is signiûcantly simpler, and thus cheaper, than the administration of many small disk stacks. However, the administrator must plan what he is doing more precisely when working with large disk subsystems. This chapter describes the functions of such modern disk subsystems.

This chapter begins with an overview of the internal structure of a disk subsystem. We then go on to consider the hard disks used inside the system and the configuration options for the internal I/O channels. The controller represents the control

Inside this Chapter

centre of a disk subsystem. Disk subsystems without controllers are called JBODs (Just a Bunch of Disks); JBODs provide only an enclosure and a common power supply for several hard disks. So-called RAID controllers bring together several physical hard disks to form virtual hard disks that are faster and more fault-tolerant than individual physical hard disks. Some RAID con-trollers use a cache to further accelerate write and read access to the server.

3.1. COMPONENT ARCHITECTURE OF INTELLIGENT DISK SUBSYSTEMS

In contrast to a file server, a disk subsystem can be visualized as a hard disk server. Servers are connected to the connection port of the disk subsystem using standard I/O techniques such as SCSI, Fibre Channel or iSCSI and can thus use the storage capacity that the disk subsystem provides (Fig. 3.1). The internal structure of the disk subsystem is completely hidden from the server, which sees only the hard disks that the disk subsystem provides to the server.

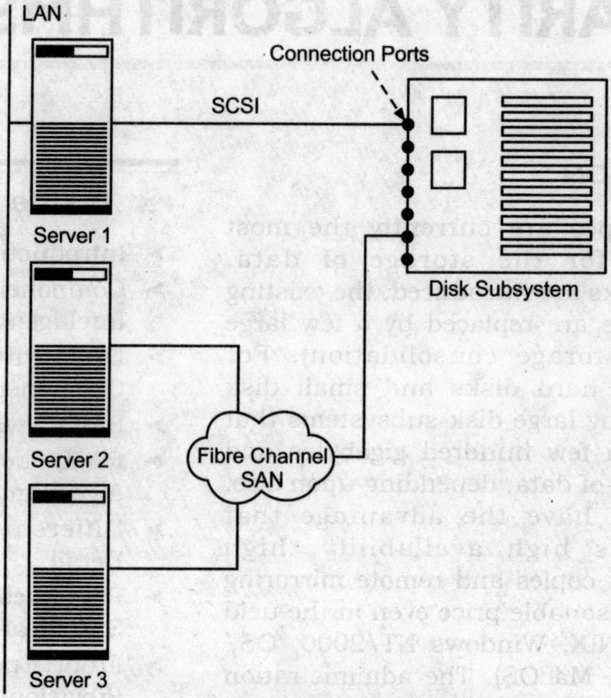

Fig. 3.1. Servers are connected to a disk subsystem using standard I/O techniques. The figure shows a server that is connected by SCSI. Two others are connected by Fibre Channel SAN

The connection ports are extended to the hard disks of the disk subsystem by means of internal I/O channels (Fig. 3.2). In most disk subsystems there is a controller between the connection ports and the hard disks. The controller can significantly increase the data availability and data access performance with the aid of a so-called RAID procedure (RAID = Redundant Array of Independent Disks). Furthermore, some controllers realize the copying services instant copy and remote mirroring and further additional services. The controller uses a cache in an attempt to accelerate read and write accesses to the server.

All sizes of disk subsystems are available. Small disk subsystems have one to two connections for servers or storage networks, six to eight hard disks and—depending upon disk capacity—a storage capacity of around 500 gigabytes. Large disk subsystems have several ten ports for servers or storage networks, redundant controllers and several internal I/O channels. Connection via a storage network means that a significantly greate number of servers can access the disk subsystem. Large disk subsystems can store several ten terabytes of data and—depending upon the manufacturer—weigh a few tons. The dimensions of a large disk subsystem are comparable with those of a wardrobe.

Figure 3.2 shows a simplified schematic representation. The architecture of real disk subsystems is more complex and varies greatly. Ultimately, however, it will always include the components shown in Fig. 3.3. The simplified representation in Fig. 3.2 provides a sufficient basis for the further discussion in the book.

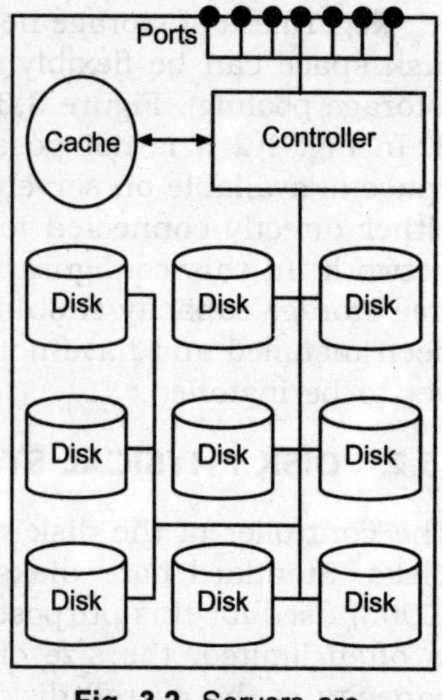

Fig. 3.2. Servers are connected to a disk subsystem via the ports. Internally, the disk subsystem consists of hard disks, a

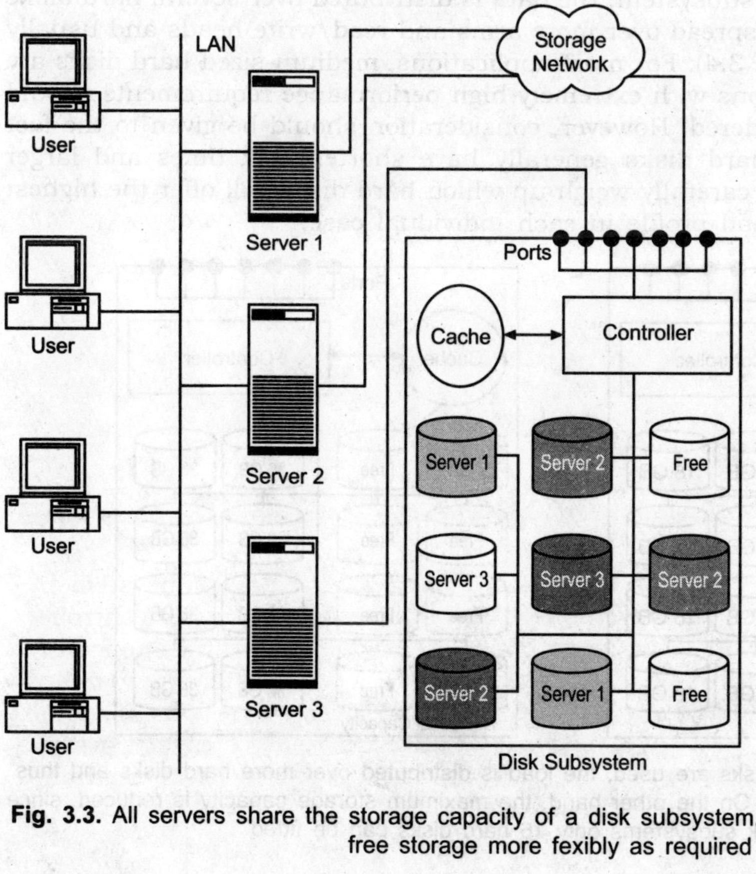

Fig. 3.3. All servers share the storage capacity of a disk subsystem. Each server can be assigned free storage more fexibly as required

Regardless of storage networks, most disk subsystems have the advantage that free disk space can be flexibly assigned to each server connected to the disk subsystem (storage pooling). Figure 3.3 refers back once again to the example of Fig. 1.2 on page 3. In Fig. 1.2 it is not possible to assign more storage to server 2, even though free space is available on servers 1 and 3. In Fig. 3.3 this is not a problem. All servers are either directly connected to the disk subsystem or indirectly connected via a storage network. In this configuration each server can be assigned free storage. Incidentally, free storage capacity should be understood to mean both hard disks that have already been installed and have not yet been used and also free slots for hard disks that have yet to be installed.

3.2. DISK PHYSICAL STRUCTURE COMPONENTS

The controller of the disk subsystem must ultimately store all data on physical hard disks. Standard hard disks that range in size from 18 GB to 250 GB are currently (2003) used for this purpose. Since the maximum number hard disks that can be used is often limited, the size of the hard disk used gives an indication of the maximum capacity of the overall disk subsystem.

When selecting the size of the internal physical hard disks it is necessary to weigh the requirements of maximum performance against those of the maximum capacity of the overall system. With regard to performance it is often beneficial to use smaller hard disks at the expense of the maximum capacity: given the same capacity, if more hard disks are available in a disk subsystem, the data is distributed over several hard disks and thus the overall load is spread over more arms and read/write heads and usually over more I/O channels (Fig. 3.4). For most applications, medium-sized hard disks are sufficient. Only for applications with extremely high performance requirements should smaller hard disks be considered. However, consideration should be given to the fact that more modern, larger hard disks generally have shorter seek times and larger caches, so it is necessary to carefully weigh up which hard disks will offer the highest performance for a certain load profile in each individual case.

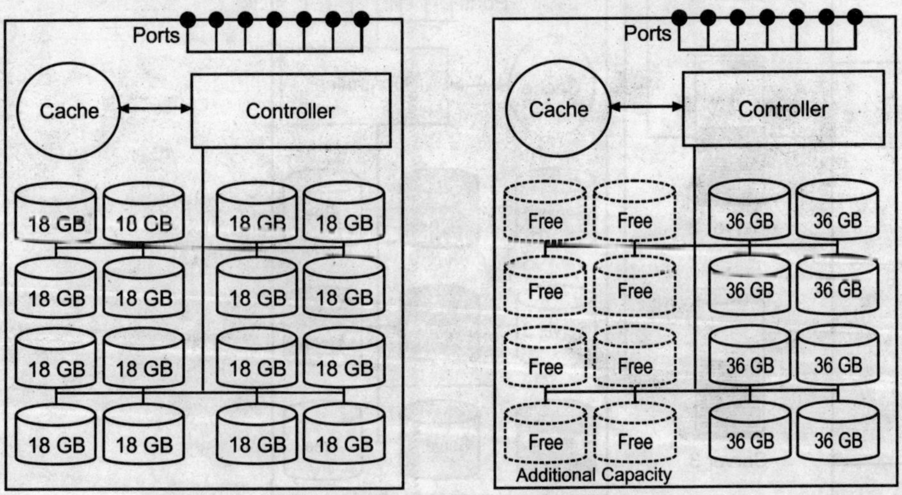

Fig. 3.4. If small internal hard disks are used, the load is distributed over more hard disks and thus over more read and write heads. On the other hand, the maximum storage capacity is reduced, since in both disk subsystems only 16 hard disks can be fitted

Standard I/O techniques such as SCSI and Fibre Channel, to an increasing degree SATA (Serial ATA) and sometimes also SSA (Serial Storage Architecture) are very often used for the internal I/O channels between connection ports and controller and between controller and internal hard disks. Sometimes, however, proprietary – i.e. manufacturer-speciûc – I/O technologies are used. Regardless of the I/O technology used, the I/O channels can be designed with built-in redundancy in order to increase the fault-tolerance of the disk subsystem. The following cases can be differentiated here:

- **Active:** In active cabling the individual physical hard disks are only connected via one I/O channel Fig. 3.5, left. If this access path fails, then it is no longer possible to access the data.
- **Active/passive:** In active/passive cabling the individual hard disks are connected via two I/O channels Fig. 3.5, right. In normal operation the controller communicates with the hard disks via the ûrst I/O channel and the second I/O channel is not used. In the event of the failure of the ûrst I/O channel, the disk subsystem switches from the first to the second I/O channel.

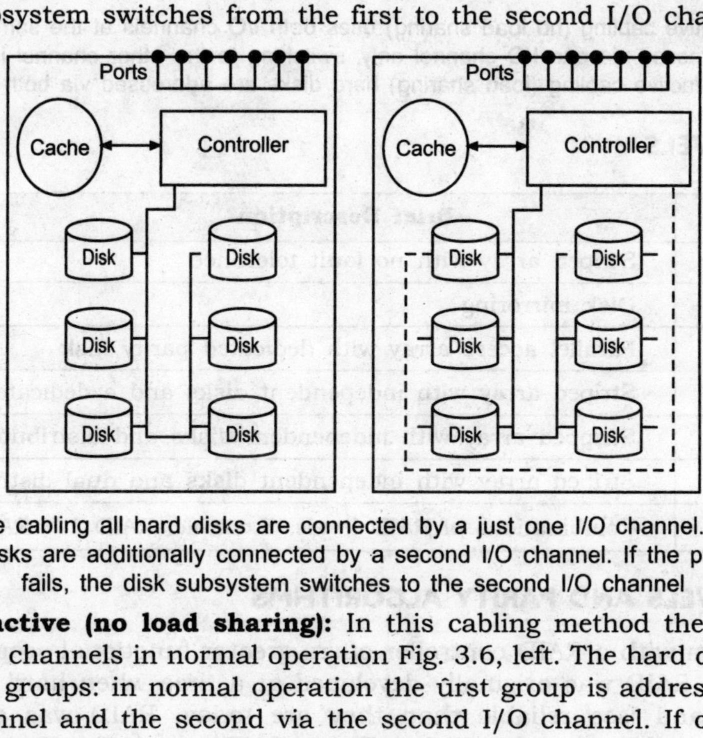

Fig. 3.5. In active cabling all hard disks are connected by a just one I/O channel. In active/passive cabling all hard disks are additionally connected by a second I/O channel. If the primary I/O channel fails, the disk subsystem switches to the second I/O channel

- **Active/active (no load sharing):** In this cabling method the controller uses both I/O channels in normal operation Fig. 3.6, left. The hard disks are divided into two groups: in normal operation the ûrst group is addressed via the first I/O channel and the second via the second I/O channel. If one I/O channel fails, both groups are addressed via the other I/O channel.
- **Active/active (load sharing):** In this approach all hard disks are addressed via both I/O channels in normal operation Fig. 3.6, right. The controller divides the load dynamically between the two I/O channels so that the available hardware can be optimally utilized. If one I/O channel fails, then the communication goes through the other channel only.

Active cabling is the simplest and thus also the cheapest to realize but offers no protection against failure. Active/passive cabling is the minimum needed to protect against failure, whereas active/active cabling with load sharing best utilizes the underlying hardware.

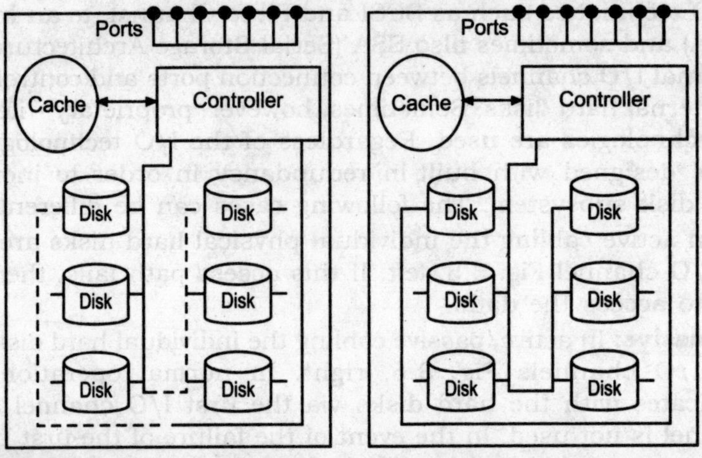

Fig. 3.6. Active/active cabling (no load sharing) uses both I/O channels at the same time. However, each disk is addressed via one I/O channel only, switching to the other channel in the event of a fault. In active/active cabling (load sharing) hard disks are addressed via both I/O channels

3.3. RAID LEVELS

Levels	Brief Description
RAID 0	Striped array with no fault tolerance
RAID 1	Disk mirroring
RAID 3	Parallel access array with dedicated parity disk
RAID 4	Striped array with independent disks and a dedicated parity disk
RAID 5	Stripped array with independent disks and distributed parity
RAID 6	Striped array with independent disks and dual distributed parity
Nested	Combinations of RAID levels. Example: RAID 1 + RAID 0

3.4. RAID LEVELS AND PARITY ALGORITHMS

A disk subsystem with a RAID controller offers greater functional scope than a JBOD disk subsystem. RAID was originally developed at a time when hard disks were still very expensive and less reliable than they are today. RAID was originally called 'Redundant Array of Inexpensive Disks'. Today RAID stands for 'Redundant Array of Independent Disks'. Disk subsystems that support RAID are sometimes also called RAID arrays.

RAID has two main goals: to increase performance by striping and to increase fault- tolerance by redundancy. Striping distributes the data over several hard disks and thus distributes the load over more hardware. Redundancy means that additional information is stored so that the operation of the application itself can continue in the event of the failure of a hard disk. You cannot increase the performance of an individual hard disk any more than you can improve its fault-tolerance. Individual physical hard disks are slow and have a limited life-cycle. However, through a suitable combination of physical hard disks it is possible to significantly increase the fault-tolerance and performance of the system as a whole.

In Fig. 3.7, the computation of parity is represented as a simple arithmetic operation on the data. However, parity calculation is a *bitwise XOR* operation. Calculation of parity is a function of the RAID controller. Compared to mirroring, parity implementation considerably reduces the cost associated with data protection. Consider a RAID configuration with five disks. Four of these disks hold data, and the fifth holds parity information. Parity requires 25 percent extra disk space compared to mirroring, which requires 100 percent extra disk space. However, there are some disadvantages of using parity. Parity information is generated from data on the data disk. Therefore, parity is recalculated every time there is a change in data. This recalculation is time-consuming and affects the performance of the RAID controller.

Parity Algorithms

Parity is a method of protecting striped data from HDD failure without the cost of mirroring. An additional HDD is added to the stripe width to hold parity, a mathematical construct that allows re-creation of the missing data. Parity is a redundancy check the ensures full protection of data without maintaining a full set of duplicate data.

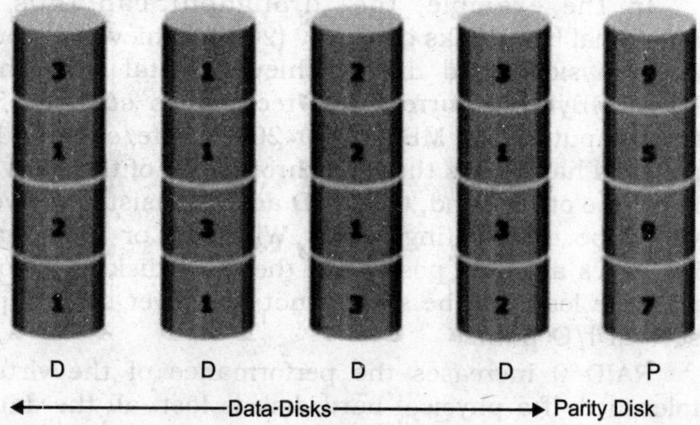

Fig. 3.7.

Parity information can be stored on separate, dedicated HDDs or distributed across all the drives in a RAID set. Figure 3.7., shows a parity RAID. The first four disks, labeled *D*, contain the data. The fifth disk, labeled *P*, stored the parity information, which in this case is the sum of the elements in each row. Now, if one of the *D*s fails, the missing value can be calculated by subtracting the sum of the rest of the elements from the parity value.

3.5. DIFFERENT RAID LEVELS IN DETAIL

RAID has developed since its original definition in 1987. Due to technical progress some RAID levels are now practically meaningless, while others have been modified or added at a later date. This section introduces the RAID levels that are currently the most significant in practice. We will not introduce RAID levels that represent manufacturer-specific variants and variants that only deviate slightly from the basic forms mentioned in the following.

3.5.1. RAID 0: Block-by-Block Striping

RAID 0 distributes the data that the server writes to the virtual hard disk onto one physical hard disk after another block-by-block (block-by-block striping). Figure 3.9 shows a RAID array with four physical hard disks. In Fig. 3.9 the server writes the blocks A, B, C, D, E, etc. onto the virtual hard disk one after the other. The RAID controller distributes the sequence of blocks onto the individual physical hard disks: it writes the first block, A, to the first physical hard disk, the second block, B, to the second physical hard disk, block C to the third and block D to the fourth. Then it begins to write to the first physical hard disk once again, writing block E to the first disk, block F to the second, and so on.

RAID 0 increases the performance of the virtual hard disk as follows: the individual hard disks can exchange data with the RAID controller via the I/O channel significantly more quickly than they can write to or read from the rotating disk. The RAID controller sends the first block, block A, to the first hard disk. This takes some time to write the block to the disk. While the first disk is writing the first block to the physical hard disk, the RAID controller is already sending the second block, block B, to the second hard disk and block C to the third hard disk. In the meantime the first two physical hard disks are still engaged in depositing their respective blocks onto the physical hard disk. If the RAID controller now sends block E to the first hard disk, then this has written block A at least partially, if not entirely, to the physical hard disk.

In the example, the throughput can thus be approximately quadrupled: individual hard disks currently (2003) achieve a throughput of around 50 MByte/s. The four physical hard disks achieve a total throughput of around 4 × 50 MByte/s H200 MByte/s. Current I/O techniques such as SCSI or Fibre Channel achieve a throughput of 160 MByte/s or 200 MByte/s. If the RAID array consisted of just three physical hard disks the total throughput of the hard disks would be the limiting factor. If, on the other hand, the RAID array consisted of five physical hard disks the I/O path would be the limiting factor. With five or more hard disks, therefore, performance increases are only possible if the hard disks are connected to different I/O paths so that the load can be striped not only over several physical hard disks, but also over several I/O paths.

RAID 0 increases the performance of the virtual hard disk, but not its fault-tolerance. If a physical hard disk is lost, all the data on the virtual hard disk is lost. To be precise, therefore, the 'R' for 'Redundant' in RAID is incorrect in the case of RAID 0, with 'RAID 0' standing instead for 'zero redundancy'.

3.5.2. RAID 1: Block-by-Block Mirroring

In contrast to RAID 0, in RAID 1 fault-tolerance is of primary importance. The basic form of RAID 1 brings together two physical hard disks to form a virtual hard disk by mirroring the data on the two physical hard disks. If the server writes a block to the virtual hard disk, the RAID controller writes this block to both physical hard disks Fig. 3.8. The individual copies are also called mirrors. Normally, two or sometimes three copies of the data are kept (three-way mirror).

In a normal operation with pure RAID 1, performance increases are only possible in read operations. After all, when reading the data the load can be divided between the two disks. However, this gain is very low in comparison to RAID 0. When writing with RAID 1 it tends to be the case that reductions in performance may even have to be taken into account. This is because the RAID controller has to send the data to both hard disks.

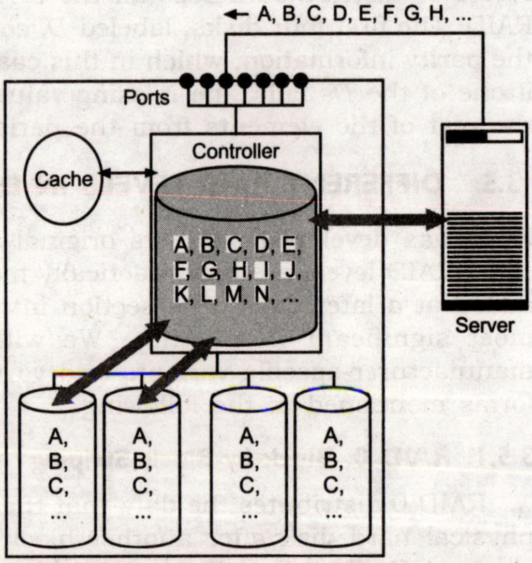

Fig. 3.8. RAID 1 (mirroring): as in all RAID levels, the server sees only the virtual hard disk. The RAID controller duplicates each of the server's write operations onto two physical hard disks. After the failure of one physical hard disk the data can still be read from the other disk

This disadvantage can be disregarded for an individual write operation, since the capacity of the I/O channel is significantly higher than the maximum write speed of the two hard disks put together. However, the I/O channel is under twice the load, which hinders other data traffic using the I/O channel at the same time.

3.5.3. RAID 0+1/RAID 10: Striping and Mirroring Combined

The problem with RAID 0 and RAID 1 is that they increase either performance (RAID 0) or fault-tolerance (RAID 1). However, it would be nice to have both performance and fault-tolerance. This is where RAID 0+1 and RAID 10 come into play. These two RAID levels combine the ideas of RAID 0 and RAID 1.

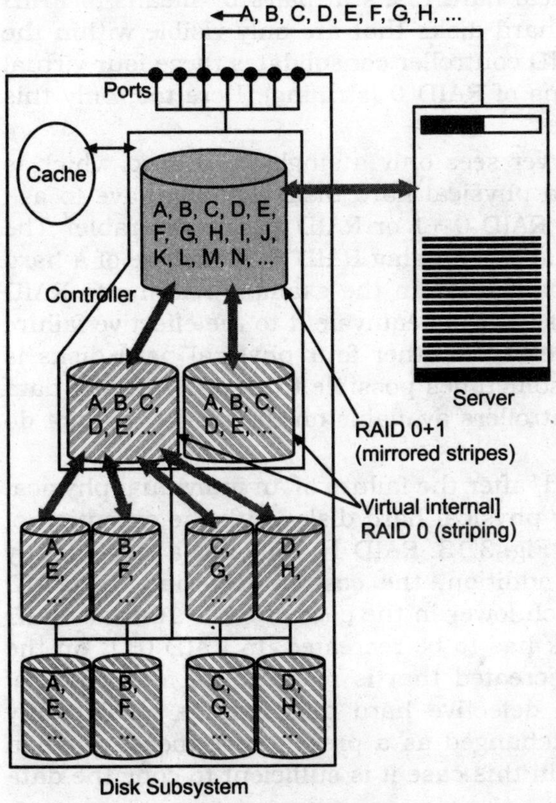

Fig. 3.9. RAID 0+1 (mirrored stripes): as in all RAID levels, the server sees only the virtual hard disk. Internally, the RAID controller realizes the virtual disk in two stages: in the ûrst stage it brings together every four physical hard disks into one virtual hard disk that is only visible within the RAID controller by means of RAID 0 (striping). In the second stage it consolidates these two virtual hard disks by means of RAID 1 (mirroring) to form the hard disk that is visible to the server

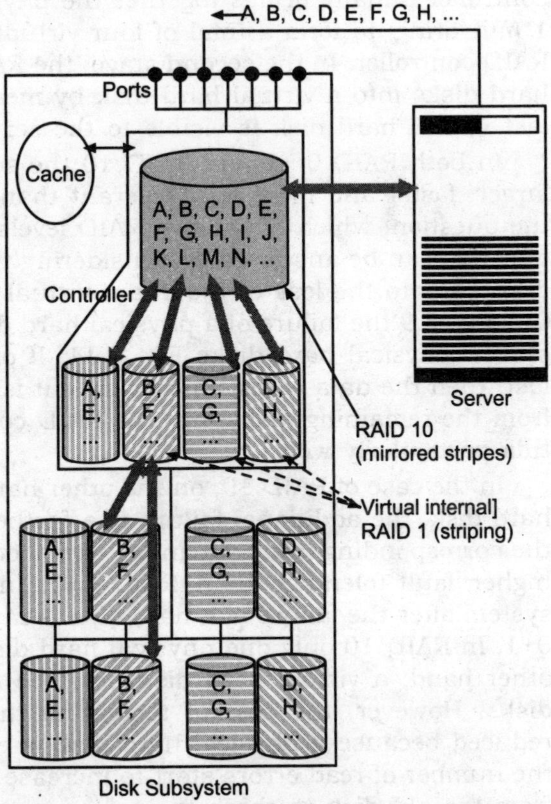

Fig. 3.10. RAID 10 (striped mirrors): as in all RAID levels, the server sees only the virtual hard disk. Here too, we proceed in two stages. The sequence of striping and mirroring is reversed in relation to RAID 0+1. In the first stage the controller links every two physical hard disks by means of RAID 1 (mirroring) to a virtual hard disk, which it uniûes by means of RAID 0 (striping) in the second stage to form the hard disk that is visible to the server

RAID 0 +1 and RAID 10 each represent a two-stage virtualization hierarchy. Fig. 3.9 shows the principle behind RAID 0+1 (mirrored stripes). In the example, eight physical hard disks are used. The RAID controller initially brings together each four physical hard disks to form a total of two virtual hard disks that are only visible within the RAID controller by means of RAID 0 (striping). In the second level, it consolidates these two virtual hard disks into a single virtual hard disk by means of RAID 1 (mirroring); only this virtual hard disk is visible to the server.

In RAID 10 (striped mirrors) the sequence of RAID 0 (striping) and RAID 1 (mirroring) is reversed in relation to RAID 0+1 (mirrored stripes). Figure 3.10 shows the principle underlying RAID 10 based again on eight physical hard disks. In RAID 10 the RAID controller initially brings together the physical hard disks in pairs by means of RAID 1 (mirroring) to form a total of four virtual hard disks that are only visible within the RAID controller. In the second stage, the RAID controller consolidates these four virtual hard disks into a virtual hard disk by means of RAID 0 (striping). Here too, only this last virtual hard disk is visible to the server.

In both RAID 0 +1 and RAID 10 the server sees only a single hard disk, which is larger, faster and more fault-tolerant than a physical hard disk. We now have to ask the question: which of the two RAID levels, RAID 0 +1 or RAID 10, is preferable? The question can be answered by considering that when using RAID 0 the failure of a hard disk leads to the loss of the entire virtual hard disk. In the example relating to RAID 0+1 Fig. 3.9 the failure of a physical hard disk is thus equivalent to the effective failure of four physical hard disks Fig. 3.11. If one of the other four physical hard disks is lost, then the data is lost. In principle it is sometimes possible to reconstruct the data from the remaining disks, but the RAID controllers available on the market cannot do this particularly well.

In the case of RAID 10, on the other hand, after the failure of an individual physical hard disk, the additional failure of a further physical hard disk—with the exception of the corresponding mirror—can be withstood Fig. 3.13. RAID 10 thus has a significantly higher fault-tolerance than RAID 0+1. In addition, the cost of restoring the RAID system after the failure of a hard disk is much lower in the case of RAID 10 than RAID 0+1. In RAID 10 only one physical hard disk has to be recreated. In RAID 0+1, on the other hand, a virtual hard disk must be recreated that is made up of four physical disks. However, the cost of recreating the defective hard disk can be significantly reduced because a physical hard disk is exchanged as a preventative measure when the number of read errors start to increase. In this case it is sufficient to copy the data from the old disk to the new.

However, things look different if the performance of RAID 0+1 is compared with the performance of RAID 10. In we discuss a case study in which the use of RAID 0 +1 is advantageous.

With regard to RAID 0+1 and RAID 10 it should be borne in mind that the two RAID procedures are often confused. Therefore, the answer 'We use RAID 10!' or 'We use RAID 0+1' does not always provide the necessary clarity. In discussions it is better to ask if mirroring takes place first and the mirror is then striped or if striping takes place first and the stripes are then mirrored.

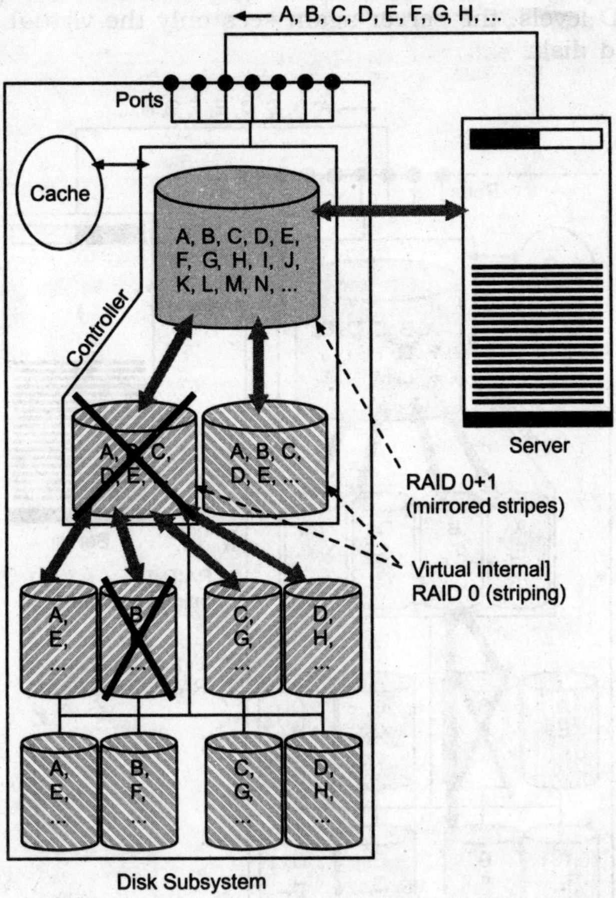

Fig. 3.11. The consequences of the failure of a physical hard disk in RAID 0+1 (mirrored stripes) are relatively high in comparison to RAID 10 (striped mirrors). The failure of a physical hard disk brings about the failure of the corresponding internal RAID 0 disk, so that in effect half of the physical hard disks have failed. The restoration of the data from the failed disk is expensive

3.5.4. RAID 4 and RAID 5: parity instead of mirroring

RAID 10 provides excellent performance at a high level of fault-tolerance. The problem with this is that mirroring using RAID 1 means that all data is written to the physical hard disk twice. RAID 10 thus doubles the required storage capacity.

The idea of RAID 4 and RAID 5 is to replace all mirror disks of RAID 10 with a single parity hard disk. Figure 3.14 shows the principle of RAID 4 based upon five physical hard disks. The server again writes the blocks A, B, C, D, E, etc. to the virtual hard disk sequentially. The RAID controller stripes the data blocks over the first four physical hard disks. Instead of mirroring all data onto the further four physical hard disks, as in RAID 10, the RAID controller calculates a parity block for every four blocks and writes this onto the fifth physical hard disk. For example, the RAID controller calculates the parity block PABCD for the blocks A, B, C and D. If one of the four data disks fails, the RAID controller can reconstruct the data of the defective disks using the three other data disks and the parity disk. In comparison to the examples in

Figures 3.10 (RAID 0+1) and 3.11 (RAID 10), RAID 4 saves three physical hard disks. As in all other RAID levels, the server again sees only the virtual disk, as if it were a single physical hard disk.

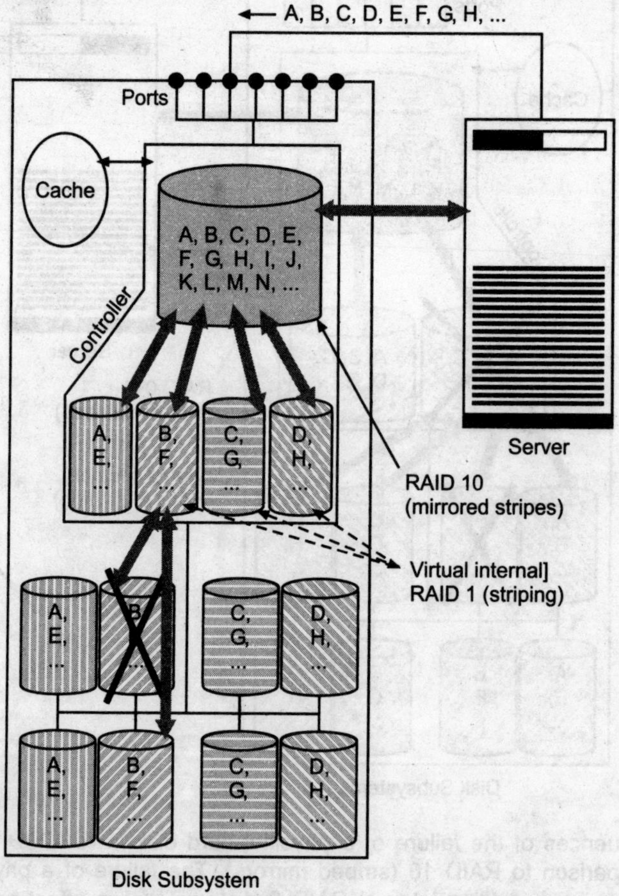

Fig. 3.12. In RAID 10 (striped mirrors) the consequences of the failure of a physical hard disk are not as serious as in RAID 0 +1 (mirrored stripes). All virtual hard disks remain intact. The restoration of the data from the failed hard disk is simple

From a mathematical point of view the parity block is calculated with the aid of the logical XOR operator (Exclusive OR). In the example from Fig. 3.13, for example, the equation PABCD¯A XOR B XOR C XOR D applies.

The space saving offered by RAID 4 and RAID 5, which remains to be discussed, comes at a price in relation to RAID 10. Changing a data block changes the value of the associated parity block. This means that each write operation to the virtual hard disk requires (1) the physical writing of the data block, (2) the recalculation of the parity block and (3) the physical writing of the newly calculated parity block. This extra cost for write operations in RAID 4 and RAID 5 is called the write penalty of RAID 4 or the write penalty of RAID 5.

The cost for the recalculation of the parity block is relatively low due to the mathematical properties of the XOR operator. If the block A is overwritten by block A and is the difference between the old and new data block, then A XOR A. The new parity block *P* can now simply be calculated from the old parity block P and, i.e. *P*˜ = P XOR. Proof of this property can be found in Appendix A. Therefore, if PABCD is the parity block for the data blocks A, B, C and D, then after the data block A has been changed, the new parity block can be calculated without knowing the remaining blocks *B*, *C* and *D*. However, the old block A must be read in before overwriting the physical hard disk in the controller, so that this can calculate the difference.

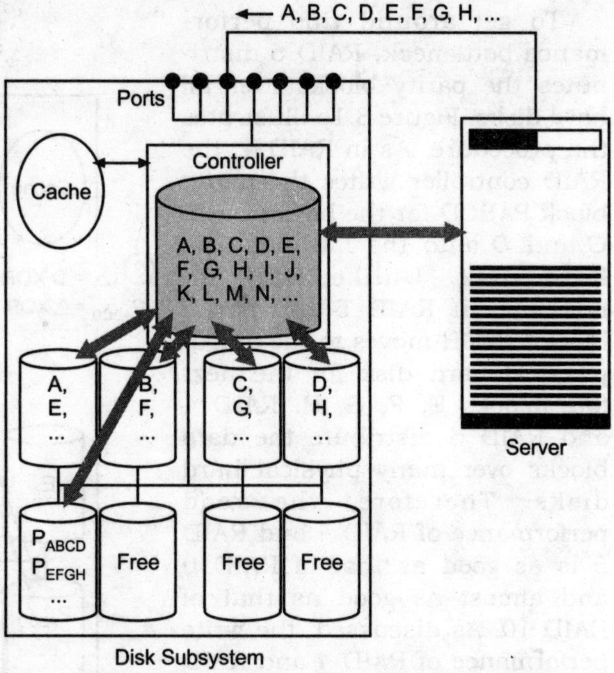

Fig. 3.13. RAID 4 (parity disk) is designed to reduce the storage requirement of RAID 0+1 and RAID 10. In the example, the data blocks are distributed over four physical hard disks by means of RAID 0 (striping). Instead of mirroring all data once again, only a parity block is stored for each four blocks

When processing write commands for RAID 4 and RAID 5 arrays, RAID controllers use the above-mentioned mathematical properties of the XOR operation for the recalculation of the parity block. Figure 3.14 shows a server that changes block D on the virtual hard disk. The RAID controller reads the data block and the associated parity block from the disk in question into its cache. Then it uses the XOR operation to calculate the difference between the old and the new parity block, i.e. = D XOR D, and from this the new parity block P˜ ABCD by means of P˜ ABCD=PABCD^{XOR}. Therefore, it is not necessary to read in all four associated data blocks to recalculate the parity block. To conclude the write operation to the virtual hard disk, the RAID controller writes the new data block and the recalculated parity block onto the physical hard disks in question.

Good RAID 4 and RAID 5 implementations are capable of reducing the write penalty even further for certain load proûles. For example, if large data quantities are written sequentially, then the RAID controller can calculate the parity blocks from the data flow without reading the old parity block from the disk. If, for example, the blocks E, F, G and H in Fig. 3.13 are written in one go, then the controller can calculate the parity block PEFGH from them and overwrite this without having previously read in the old value. Likewise, a RAID controller with a suitably large cache can hold frequently changed parity blocks in the cache after writing to the disk, so that the next time one of the data blocks in question is changed there is no need to read in the parity block. In both cases the I/O load is now lower than in the case of RAID 10. In the example only five physical blocks now need to be written instead of eight as is the case with RAID 10.

RAID 4 saves all parity blocks onto a single physical hard disk. For example, in Fig. 3.13 this means that the write operations for the data blocks are distributed over four physical hard disks. However, the parity disk has to handle the same number of write operations all on its own. Therefore, the parity disk become the performance bottleneck of RAID 4 if there are a high number of write operations.

To get around this performance bottleneck, RAID 5 distributes the parity blocks over all hard disks. Figure 3.16 illustrates the procedure. As in RAID 4, the RAID controller writes the parity block PABCD for the blocks *A*, *B*, *C* and *D* onto the fifth physical hard disk. Unlike RAID 4, however, in RAID 5 the parity block PEFGH moves to the fourth physical hard disk for the next four blocks E, F, G, H. RAID 4 and RAID 5 distribute the data blocks over many physical hard disks. Therefore, the read performance of RAID 4 and RAID 5 is as good as that of RAID 0 and almost as good as that of RAID 10. As discussed, the write performance of RAID 4 and RAID 5 suffers from the write penalty; in RAID 4 there is an additional bottleneck caused by the parity disk. Therefore, RAID 4 is seldom used in practice because RAID 5 accomplishes more than RAID 4 with the same technical resources.

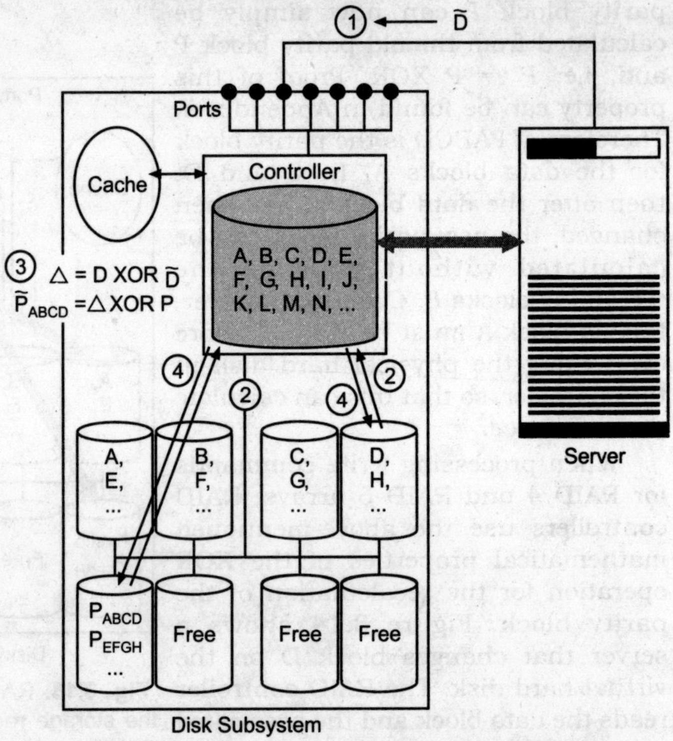

Fig 3.14. Write penalty of RAID 4 and RAID 5: the server writes a changed data block (1). The RAID controller reads in the old data block and the associated old parity block (2) and calculates the new parity block (3). Finally it writes the new data block and the new parity block onto the physical hard disk in question (4)

RAID 4 and RAID 5 can withstand the failure of a physical hard disk. The use of parity blocks means that the data on the defective hard disk can be restored with the aid of the other hard disk. In contrast to RAID 10, the failure of a second physical hard disk always leads to data loss. Some RAID 5 variants get around this by simply keeping a second parity disk so that the data is doubly protected.

In RAID 4 and RAID 5 the restoration of a defective physical hard disk is significantly more expensive than is the case for RAID 1 and RAID 10. In the latter two RAID levels only the mirror of the defective disk needs to be copied to the replaced disk. In RAID 4 and RAID 5, on the other hand, the RAID controller has to read the data from all disks, use this to recalculate the lost data blocks and parity blocks, and then write these blocks to the replacement disk. As in RAID 0+1 this high cost can be avoided by replacing a physical hard disk as a precaution as soon as the rate of read errors increases. If this is done, it is sufficient to copy the data from the hard disk to be replaced onto the new hard disk.

If the fifth physical hard disk has to be restored in the examples from Fig. 3.14 (RAID 4) and Fig. 3.15 (RAID 5), the RAID controller must first read the blocks A, B, C and D from the physical hard disks, recalculate the parity block PABCD and then write to the exchanged physical hard disk. If a data block has to be restored, only the calculation rule changes. If, in the example, the third physical hard disk is to be recreated, the controller would first have to read in the blocks A, B, D and PABCD, use these to reconstruct block C and write this to the replaced disk.

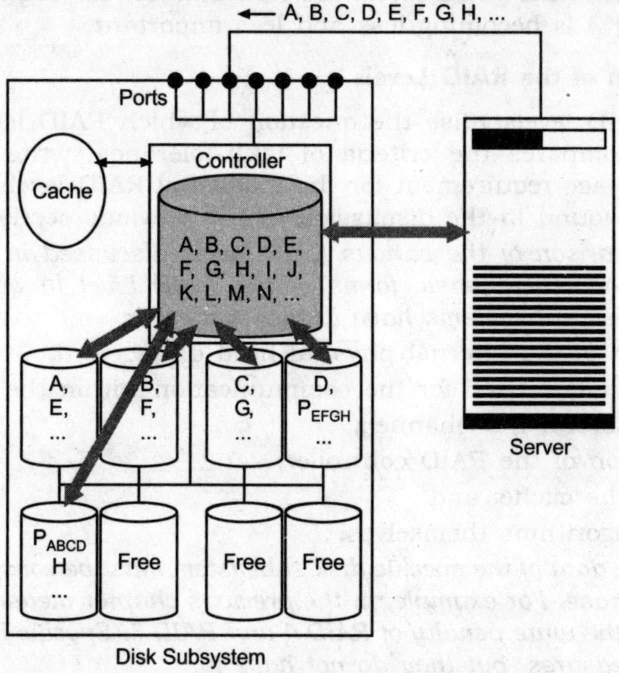

Fig. 3.15. RAID 5 (striped parity): in RAID 4 each write access by the server is associated with a write operation to the parity disk for the comparison of parity information. RAID 5 distributes the load of the parity disk over all physical hard disks

3.5.5. RAID 2 and RAID 3

When introducing the RAID levels we are sometimes asked: 'and what about RAID 2 and RAID 3?'. The early work on RAID began at a time when disks were not yet very reliable: bit errors were possible that could lead to a written 'one' being read as 'zero' or a written 'zero' being read as 'one'. In RAID 2 the Hamming code is used, so that redundant information is stored in addition to the actual data. This additional data permits the recognition of read errors and to some degree also makes it possible to correct them.

Today, comparable functions are performed by the controller of each individual hard disk, which means that RAID 2 no longer has any practical significance.

Like RAID 4 or RAID 5, RAID 3 stores parity data. RAID 3 distributes the data of a block amongst all the disks of the RAID 3 system so that, in contrast to RAID 4 or RAID 5, all disks are involved in every read or write access. RAID 3 only permits the reading and writing of whole blocks, thus dispensing with the write penalty that occurs in RAID 4 and RAID 5. The writing of individual blocks of a parity group is thus not possible.

In addition, in RAID 3 the rotation of the individual hard disks is synchronized so that the data of a block can truly be written simultaneously. RAID 3 was for a long time called the recommended RAID level for sequential write and read profiles such as data mining and video processing. Current hard disks come with a large cache of their own, which means that they can temporarily store the data of an entire track, and they have significantly higher rotation speeds than the hard disks of the past. As a result of these innovations, other RAID levels are now suitable for sequential load profiles, meaning that RAID 3 is becoming less and less important.

3.5.6. A Comparison of the RAID Levels

The various RAID levels raise the question of which RAID level should be used when. Table 3.1 compares the criteria of fault-tolerance, write performance, read performance and space requirement for the individual RAID levels. The evaluation of the criteria can be found in the discussion in the previous sections.

Note: *The comparison of the various RAID levels discussed in this section is only applicable to the theoretical basic forms of the RAID level in question. In practice, manufacturers of disk subsystems have design options in*

- the selection of the internal physical hard disks;
- the I/O technique used for the communication within the disk subsystem;
- the use of several I/O channels;
- the realization of the RAID controller;
- the size of the cache; and
- the cache algorithms themselves.

The performance data of the speciûc disk subsystem must be considered very carefully for each individual case. For example, in the previous chapter measures were discussed that greatly reduce the write penalty of RAID 4 and RAID 5. Speciûc RAID controllers may implement these measures, but they do not have to.

Table 3.1. The table compares the theoretical basic forms of the various RAID levels. In practice there are very marked differences in the quality of the implementation of RAID controllers

RAID level	Fault -tolerance	Read performance	Write performance	Space requirement
RAID 0	none	good	very good	minimal
RAID 1	high	poor	poor	high
RAID 10	very high	very hood	good	high
RAID 4	high	good	very very good	low
RAID 5	high	good	very poor	low

Subject to the above warning, RAID 0 is the choice for applications for which the maximum write performance is more important than protection against the failure of a disk. Examples are the storage of multimedia data for film and video production and the recording of physical experiments in which the entire series of measurements has no value if all measured values cannot be recorded. In this case it is more beneficial to record all of the measured data on a RAID 0 array first and then copy it after the experiment, for example on a RAID 5 array. In databases, RAID 0 is used as a fast store for segments in which intermediate results for complex requests are to be temporarily stored. However, as a rule hard disks tend to fail at the most inconvenient moment so database administrators only use RAID 0 if it is absolutely necessary, even for temporary data.

With RAID 1, performance and capacity are limited because only two physical hard disks are used. RAID 1 is therefore a good choice for small databases for which the configuration of a virtual RAID 5 or RAID 10 disk would be too large. A further important field of application for RAID 1 is in combination with RAID 0.

RAID 10 is used in situations where high write performance and high fault-tolerance are called for. For a long time it was recommended that database log files be stored on RAID 10. Databases record all changes in log files so this application has a high write component. After a system crash the restarting of the database can only be guaranteed if all log files are fully available. Manufacturers of storage systems disagree as to whether this recommendation is still valid as there are now fast RAID 4 and RAID 5 implementations.

RAID 4 and RAID 5 save disk space at the expense of a poorer write performance. For a long time the rule of thumb was to use RAID 5 where the ratio of read operations to write operations is 70: 30. At this point we wish to repeat that there are now storage systems on the market with excellent write performance that store the data internally using RAID 4 or RAID 5.

3.6. PROPERTIES, PERFORMANCE, AND SPECIFICATIONS OF DISKS

In all fields of computer systems, caches are used to speed up slow operations by operating them from the cache. Specifically in the field of disk subsystems, caches are designed to accelerate write and read accesses to physical hard disks. In this connection we can differentiate between two types of cache: (1) cache on the hard disk and (2) cache in the RAID controller. The cache in the RAID controller is subdivided into write cache and read cache.

3.6.1. Cache on the Hard Disk

Each individual hard disk comes with a very small cache. This is necessary because the transfer rate of the I/O channel to the disk controller is significantly higher than the speed at which the disk controller can write to or read from the physical hard disk. If a server or a RAID controller writes a block to a physical hard disk, the disk controller stores this in its cache. The disk controller can thus write the block to the physical hard disk in its own time whilst the I/O channel can be used for data traffic to the other hard disk.

Many RAID levels use precisely this state of affairs to increase the performance of the virtual hard disk. Read access is accelerated in a similar manner. If a server or an intermediate RAID controller wishes to read a block, it sends the address of the requested block to the hard disk controller. The I/O channel can be used for other data traffic while the hard disk controller copies the complete block from the physical hard disk into its cache at a slower data rate. The hard disk controller transfers the block from its cache to the RAID controller or to the server at the higher data rate of the I/O channel.

3.6.2. Write Cache in the Controller of the Disk Subsystem

In addition to the cache of the individual hard drives many disk subsystems come with their own cache, which in some models is gigabytes in size. As a result it can buffer much greater data quantities than the cache on the hard disk. The write cache should have a battery back-up and ideally be mirrored. The battery back-up is necessary to allow the data in the write cache to survive a power cut. A write cache with battery back-up can significantly reduce the write penalty of RAID 4 and RAID 5, particularly for sequential write access ('RAID 4 and RAID 5: parity instead of mirroring'), and smooth out load peaks.

Many applications do not write data at a continuous rate, but in batches. If a server sends several data blocks to the disk subsystem, the controller initially buffers all blocks into a write cache with a battery back-up and immediately reports back to the server that all data has been securely written to the drive. The disk subsystem then copies the data from the write cache to the slower physical hard disk in order to make space for the next write peak.

3.6.3. Read Cache in the RAID Controller

The acceleration of read operations is difûcult in comparison to the acceleration of write operations using cache. To speed up read access by the server, the disk subsystem's controller must copy the relevant data blocks from the slower physical hard disk to the fast cache before the server requests the data in question.

The problem with this is that it is very difficult for the disk subsystem's controller to work out in advance what data the server will ask for next. The controller in the disk subsystem knows neither the structure of the information stored in the data blocks nor the access pattern that an application will follow when accessing the data. Consequently, the controller can only analyse past data access and use this to extrapolate which data blocks the server will access next. In sequential read processes this prediction is comparatively simple, in the case of random access it is almost impossible. As a rule of thumb, good RAID controllers manage to provide around 40% of the requested blocks from the read cache in mixed read profiles.

The disk subsystem's controller cannot further increase the ratio of read access provided from the cache (pre-fetch hit rate), because it does not have the necessary application knowledge. Therefore, it is often worthwhile realizing a further cache within applications.

For example, after opening a file, file systems can load all blocks of the file into the main memory; the file system knows the structures that the files are stored in. File systems can thus achieve a prefect hit rate of 100%. However, it is impossible to know whether the expense for the storage of the blocks is worthwhile in an individual case, since the application may not actually request further blocks of the file.

3.7. FRONT END TO HOST STORAGE PROVISIONING

The front end provides the interface between the storage system and the host. It consists of two components: front-end ports and front-end controllers. The *front-end ports* enable hosts to connect to the intelligent storage system. Each front-end port has processing logic that executes the appropriate transport protocol, such as SCSI, Fibre Channel, or iSCSI, for storage connections. Redundant ports are provided on the front end for high availability.

Front-end controllers route data to and from cache via the internal data bus. When cache receives write data, the controller sends an acknowledgment message back to the host. Controllers optimize I/O processing by using command queuing algorithms.

Front-End Command Queuing

Command queuing is a technique implemented on front-end controllers. It determines the execution order of received commands and can reduce unnecessary drive head movements and improve disk performance. When a command is received for execution, the command queuing algorithms assigns tag that defines a sequence in which commands should be executed. With command queuing, multiple commands can be executed concurrently based on the organization of data on the disk, regardless of the order in which the commands were received.

The most commonly used command queuing algorithms are as follows:

- **First In First Out** (FIFO): This is the default algorithm where commands are executed in the order in which they are received (Fig. 3.16 [*a*]). There is no reordering of requests for optimization; therefore, it is inefficient in terms of performance.
- **Seek Time Optimization:** Commands are executed based on optimizing read/write head movements, which may result in reordering of commands. Without seek time optimization, the commands are executed in the order they are received. For example, as shown in Figure 3.16 (*a*), the commands are executed in the order A, B, C and D. The radial movement required by the head to execute C immediately after A is less than what would be required to execute B. With seek time optimization, the command execution sequence would be A, C, B and D.

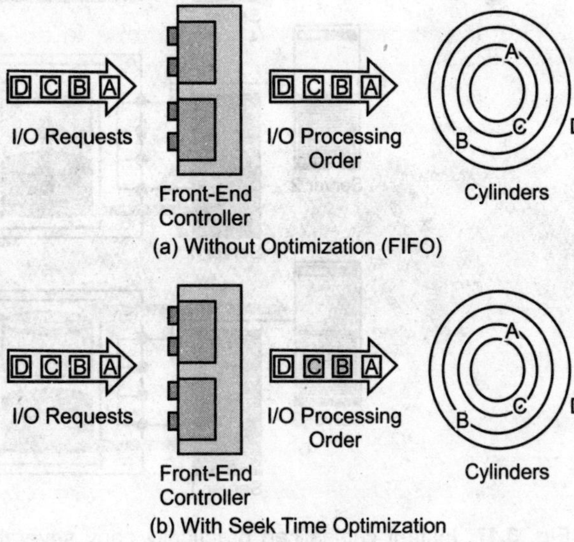

Fig. 3.16.

- **Access Time Optimization:** Commands are executed based on the combination of seek time optimization and an analysis of rotational latency for optimal performance.

Command queuing can also be implemented on disk controllers and this may further supplement the command queuing implemented on the front-end controllers. Some models of SCSI and Fibre Channel drives have command queuing implemented on their controllers.

Intelligent disk subsystems represent the third level of complexity for controllers after JBODs and RAID arrays. The controllers of intelligent disk subsystems offer additional functions over and above those offered by RAID. In the disk subsystems that are currently available on the market these functions are usually instant copies, remote mirroring and LUN masking.

3.7.1. Instant Copies

Instant copies can practically copy data sets of several terabytes within a disk subsystem in a few seconds. Virtual copying means that disk subsystems fool the attached servers into believing that they are capable of copying such large data quantities in such a short space of time. The actual copying process takes significantly longer. However, the same server, or a second server, can access the practically copied data after a few seconds (Fig. 3.17).

Instant copies are used, for example, for the generation of test data, for the back-up of data and for the generation of data copies for data mining.

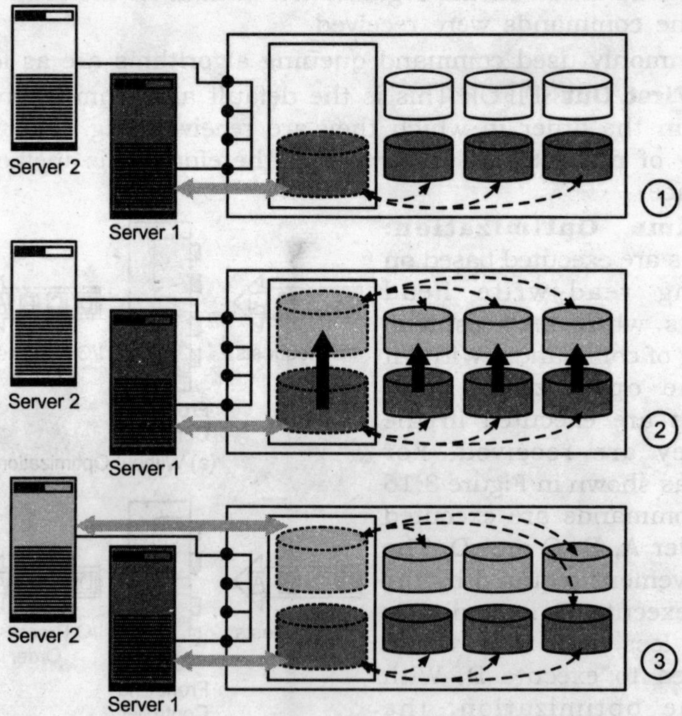

Fig. 3.17. Instant copies can practically copy several terabytes of data within a disk subsystem in a few seconds: server 1 works on the original data (1). The original data is practically copied in a few seconds (2). Then server 2 can work with the data copy, whilst server 1 continues to operate with the original data (3)

There are numerous alternative implementations for instant copies. One thing that all implementations have in common is that the pretence of being able to copy data in a matter of seconds costs resources. All realizations of instant copies require controller computing time and cache and place a load on internal I/O channels and hard disks. The different implementations of instant copy force the performance down at different times.

However, it is not possible to choose the most favourable implementation alternative depending upon the application used because real disk subsystems only ever realize one implementation alternative of instant copy.

In the following, two implementation alternatives will be discussed that function in very different ways. At one extreme the data is permanently mirrored (RAID 1 or RAID 10). Upon the copy command both mirrors are separated: the separated mirrors can then be used independently of the original. After the separation of the mirror the production data is no longer protected against the failure of a hard disk. Therefore, to increase data protection, three mirrors are often kept prior to the separation of the mirror (three-way mirror), so that the production data is always mirrored after the separation of the copy.

At the other extreme, no data at all is copied prior to the copy command, only after the instant copy has been requested. To achieve this, the controller administers two data areas, one for the original data and one for the data copy generated by means of instant copy. The controller must ensure that during write and read access operations to original data or data copies the blocks in question are written to or read from the data areas in question. In some implementations it is permissible to write to the copy, in some it is not. Some implementations copy just the blocks that have actually changed (partial copy), others copy all blocks as a background process until a complete copy of the original data has been generated (full copy).

In the following, the case differentiations of the controller will be investigated in more detail based upon the example from Fig. 3.18. We will first consider access by server 1 to the original data. Read operations are completely unproblematic; they are always served from the area of the original data. Handling write operations is trickier. If a block is changed for the first time since the generation of the instant copy, the controller must first copy the old block to the data copy area so that server 2 can continue to access the old data set. Only then may it write the changed block to the original data area. If a block that has already been changed in this manner has to be written again, it must be written to the original data area. The controller may not even back up the previous version of the block to the data copy area because otherwise the correct version of the block would be overwritten.

The case differentiations for access by server 2 to the data copy generated by means of instant copy are somewhat simpler. In this case, write operations are unproblematic: the controller always writes all blocks to the data copy area. On the other hand, for read operations it has to establish whether the block in question has already been copied or not. This determines whether it has to read the block from the original data area or read it from the data copy area and forward it to the server.

3.7.2. Remote Mirroring

Instant copies are excellently suited for the copying of data sets within disk subsystems. However, they can only be used to a limited degree for data protection. Although data copies generated using instant copy protect against application errors (accidental deletion of a file system) and logical errors (errors in the database program), they do not protect against the failure of a disk subsystem. Something as simple as a power failure can prevent access to production data and data copies for several hours. A fire in the disk subsystem would destroy original data and data copies. For data protection, therefore, the proximity of production data and data copies is fatal.

Remote mirroring offers protection against such catastrophes. Modern disk subsystems can now mirror their data, or part of their data, independently to a second disk subsystem, which is a long way away. The entire remote mirroring operation is handled by the two participating disk subsystems. Remote mirroring is invisible to application servers and does not consume their resources. However, remote mirroring requires resources in the two disk subsystems and in the I/O channel that connects the two disk subsystems together, which means that reductions in performance can sometimes make their way through to the application.

Figure 3.18 shows an application that is designed to achieve high availability using remote mirroring. The application server and the disk subsystem, plus the associated data, are installed in the primary data centre. The disk subsystem independently mirrors the application data onto the second disk subsystem that is installed 50 kilometres away in the back-up data centre by means of remote mirroring. Remote mirroring ensures that the application data in the back-up data centre is always kept up-to-date with the time interval for updating the second disk subsystem being configurable. If the disk subsystem in the primary data centre fails, the back-up application server in the back-up data centre can be started up using the data of the second disk subsystem and the operation of the application can be continued. The I/O techniques required for the connection of the two disk subsystems will be discussed in the next chapter.

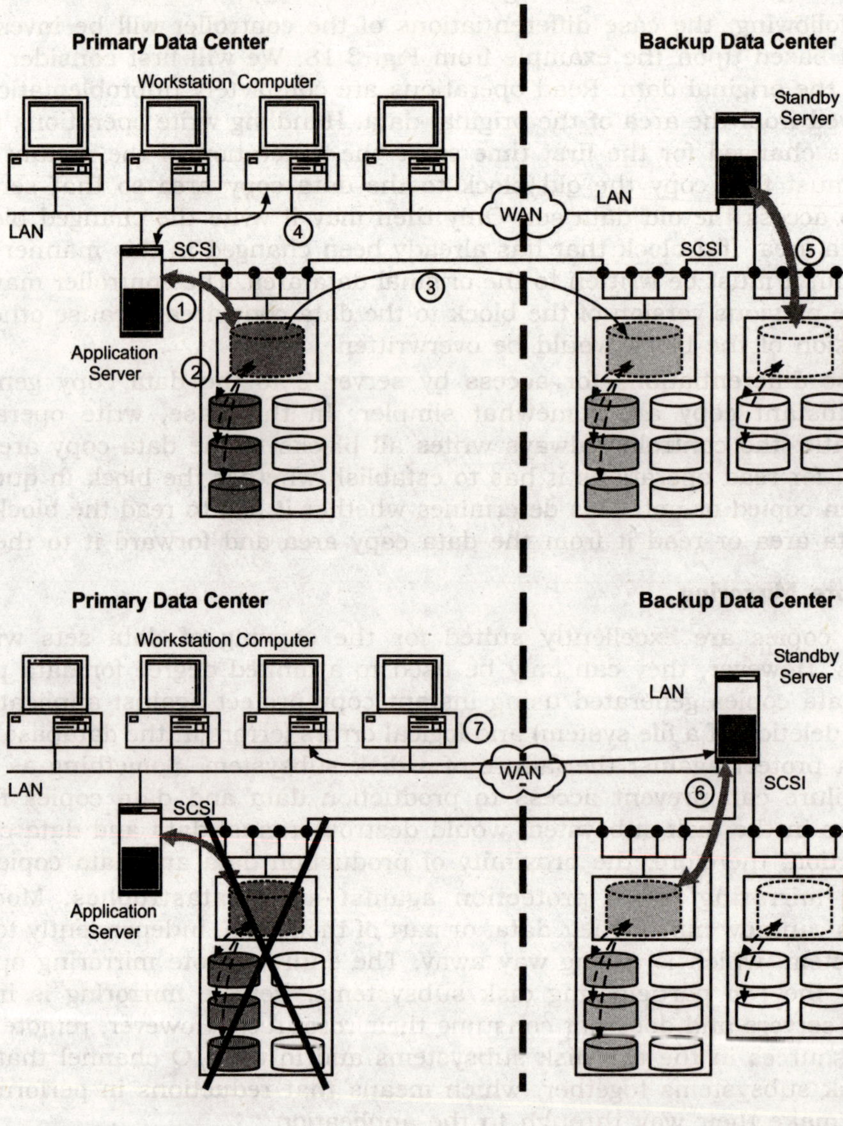

Fig. 3.18. High availability with remote mirroring: (1) The application server stores its data on a local disk subsystem. (2) The disk subsystem saves the data to several physical drives by means of RAID. (3) The local disk subsystem uses remote mirroring to mirror the data onto a second disk subsystem located in the back-up data centre. (4) Users use the application via the LAN. (5) The stand-by server in the back-up data centre is used as a test system. The test data is located on a further disk subsystem. (6) If the ûrst disk subsystem fails, the application is started up on the stand-by server using the data of the second disk subsystem. (7) Users use the application via the WAN

We can differentiate between synchronous and asynchronous remote mirroring. In synchronous remote mirroring the first disk subsystem sends the data to the second disk subsystem first before it acknowledges a server's write command. By contrast, asyn- chronous remote mirroring acknowledges a write command immediately; only then it send the copy of the block to the second disk subsystem.

Figure 3.19 illustrates the data flow of synchronous remote mirroring. The server writes block A to the first disk subsystem. This stores the block in its write cache and immediately sends it to the second disk subsystem, which also initially stores the block in its write cache. The first disk subsystem waits until the second reports that it has written the block.

The question of whether the block is still stored in the write cache of the second disk subsystem or has already been written to the hard disk is irrelevant to the first disk subsystem. It does not acknowledge to the server that the block has been written until it has received confirmation from the second disk subsystem that this has written the block.

Synchronous remote mirroring has the advantage that the copy of the data held by the second disk subsystem is always up-to-date. This means that if the first disk subsystem fails, the application can continue working with the most recent data set by utilizing the data on the second disk subsystem.

The disadvantage is that copying the data from the first disk subsystem to the second and sending the write acknowledgement back from the second to the first increases the response time of the first disk subsystem to the server. However, it is precisely this response time that determines the throughput of applications such as databases and file systems. An important factor for the response time is the signal transit time between the two disk subsystems. After all, their communication is encoded in the form of physical signals, which propagate at a certain speed. The propagation of the signals from one disk subsystem to another simply costs time. As a rule of thumb, it is worth using synchronous remote mirroring if the cable lengths from the server to the second disk subsystem via the first are a maximum of 6–10 kilometres.

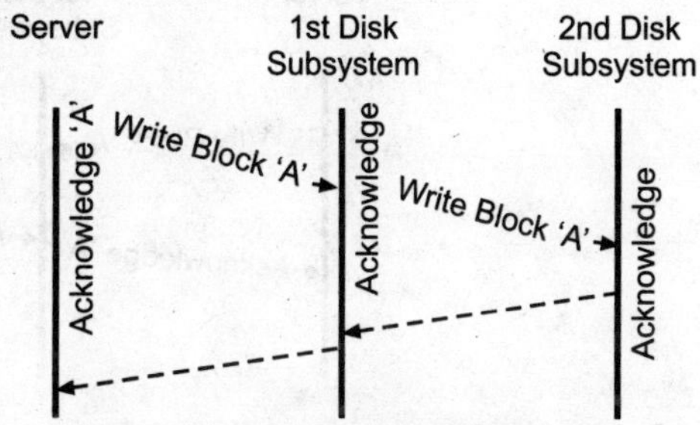

Fig. 3.19. In synchronous remote mirroring a disk subsystem does not acknowledge write operations until it has saved a block itself and received write confirmation from the second disk subsystem

If we want to mirror the data over longer distances, then we have to switch to asynchronous remote mirroring. Figure 3.20 illustrates the data flow in asynchronous remote mirroring. In this approach the first disk subsystem acknowledges the receipt of data as soon as it has been temporarily stored in the write cache. The first disk subsystem does not send the copy of the data to the second disk subsystem until later. The write confirmation of the second disk subsystem to the first is not important to the server that has written the data.

The price of the rapid response time achieved using asynchronous remote mirroring is obvious. In contrast to synchronous remote mirroring, in asynchronous remote mirroring there is no guarantee that the data on the second disk subsystem is up-to-date. This is precisely the case if the first disk subsystem has sent the write acknowledgement to the server but the block has not yet been saved to the second disk subsystem.

If we wish to mirror data over long distances but do not want to use only asynchronous remote mirroring it is necessary to use three disk subsystems (Fig. 3.21). The first two may be located just a few kilometres apart, so that synchronous remote mirroring can be used between the two. In addition, the data of the second disk subsystem is mirrored onto a third by means of asynchronous remote mirroring. However, this solution comes at a price: for most applications the cost of data protection would exceed the costs that would be incurred after data loss in the event of a catastrophe. This approach would therefore only be considered for very important applications.

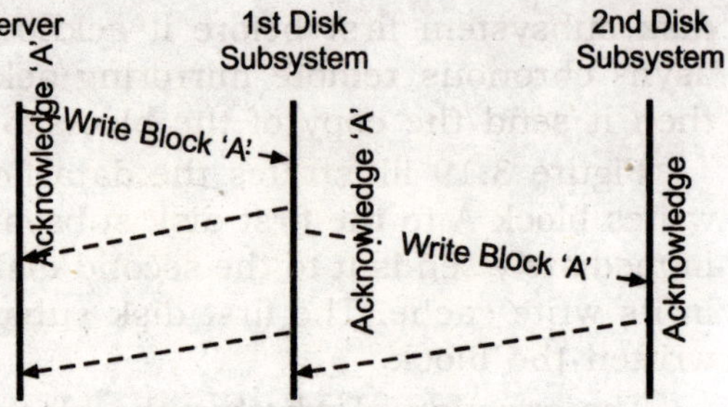

Fig. 3.20. In asynchronous remote mirroring one disk subsystem acknowledges a write operation as soon as it has saved the block itself

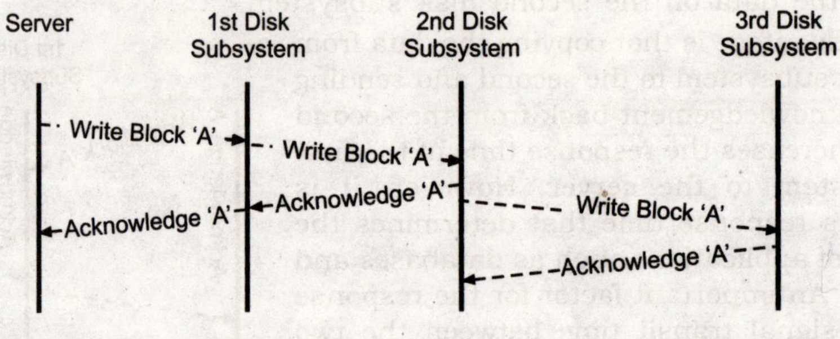

Fig. 3.21. The combination of synchronous and asynchronous remote mirroring means that rapid response times can be achieved in combination with mirroring over long distances

3.7.3. LUN Masking

So-called LUN masking brings us to the third important function – after instant copy and remote mirroring – that intelligent disk subsystems offer over and above that offered by RAID. LUN masking limits the access to the hard disks that the disk subsystem exports to the connected server. A disk subsystem makes the storage capacity of its internal physical hard disks available to servers by permitting access to individual physical hard disks, or to virtual hard disks created using RAID, via the connection ports. Based upon the SCSI protocol, all hard disks – physical and virtual – that are visible outside the disk subsystem are also known as LUN (Logical Unit Number).

Without LUN masking every server would see all hard disks that the disk subsystem provides. Figure 3.22 shows a disk subsystem without LUN masking to which three servers are connected. Each server sees all hard disks that the disk subsystem exports outwards. As a result, considerably more hard disks are visible to each server than is necessary.

In particular, on each server those hard disks that are required by applications that run on a different server are visible. This means that the individual servers must be very carefully configured. In Fig. 3.22 an erroneous formatting of the disk LUN 3 of server 1 would destroy the data of the application that runs on server 3. In addition, some operating systems are very greedy: when booting up they try to draw to them each hard disk that is written with the signature (label) of a foreign operating system.

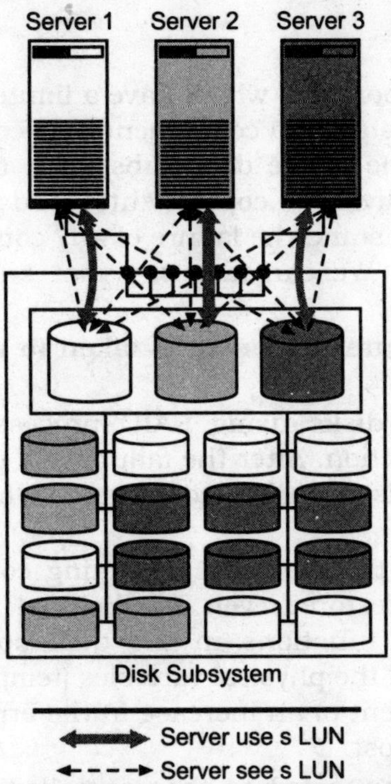

Fig. 3.22. Chaos: each server works to its own virtual hard disk. Without LUN masking each server sees all hard disks. A configuration error on server 1 can destroy the data on the other two servers. The data is thus poorly protected

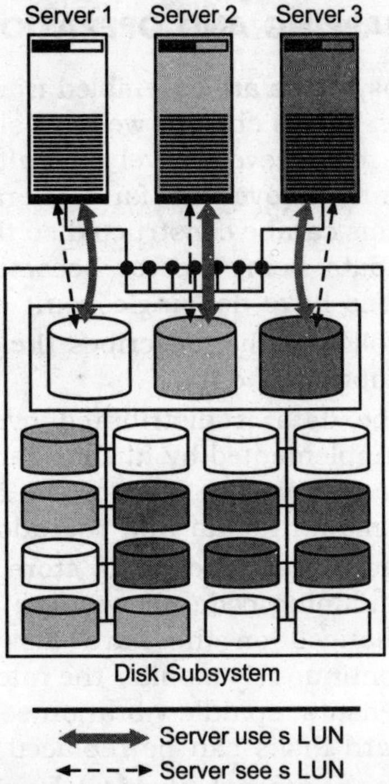

Fig. 3.23. Order: each server works to its own virtual hard disk. With LUN masking, each server sees only its own hard disks. A configuration error on server 1 can no longer destroy the data of the two other servers. The data is now protected

Without LUN masking, therefore, the use of the hard disk must be very carefully configured in the operating systems of the participating servers. LUN masking brings order to this chaos by assigning the hard disks that are externally visible to servers. As a result, it limits the visibility of exported disks within the disk subsystem. Fig. 3.24 shows how LUN masking brings order to the chaos of Fig. 3.23. Each server now sees only the hard disks that it actually requires. LUN masking thus acts as a filter between the exported hard disks and the accessing servers.

It is now no longer possible to destroy data that belongs to applications that run on another server. Configuration errors are still possible, but the consequences are no longer so devastating. Furthermore, configuration errors can now be more quickly traced since the information is bundled within the disk subsystem instead of being distributed over all servers.

We differentiate between port-based LUN masking and server-based LUN masking. Port-based LUN masking is the 'poor man's LUN masking', it is found primarily in low-end disk subsystems. In port-based LUN masking the ûlter only works using the granularity of a port. This means that all servers connected to the disk subsystem via the same port see the same disks.

Server-based LUN masking offers more flexibility. In this approach every server sees only the hard disks assigned to it, regardless of which port it is connected via or which other servers are connected via the same port.

3.8. MAPPING AND OPERATION

Disk subsystems are assembled from standard components, which have a limited fault-tolerance. In this chapter we have shown how these standard components are combined in order to achieve a level of fault-tolerance for the entire disk subsystem that lies significantly above the fault-tolerance of the individual components. Today, disk subsystems can be constructed so that they can withstand the failure of any component without data being lost or becoming inaccessible. We can also say that such disk subsystems have no 'single point of failure'.

The following list describes the individual measures that can be taken to increase the availability of data:

- The data is distributed over several hard disks using RAID processes and supplemented by further data for error correction. After the failure of a physical hard disk, the data of the defective hard disk can be reconstructed from the remaining data and the additional data.
- Individual hard disks store the data using the so-called Hamming code. The Hamming code allows data to be correctly restored even if individual bits are changed on the hard disk. Self-diagnosis functions in the disk controller continuously monitor the rate of bit errors and the physical variables (temperature sensors, spindle vibration sensors). In the event of an increase in the error rate, hard disks can be replaced before data is lost.
- Each internal physical hard disk can be connected to the controller via two internal I/O channels. If one of the two channels fails, the other can still be used.
- The controller in the disk subsystem can be realized by several controller instances. If one of the controller instances fails, one of the remaining instances takes over the tasks of the defective instance.
- Other auxiliary components such as power supplies, batteries and fans can often be duplicated so that the failure of one of the components is unimportant. When connecting the power supply it should be ensured that the various power cables are at least connected through various fuses. Ideally, the individual power cables would be supplied via different external power networks; however, in practice this is seldom realizable.
- Server and disk subsystem are connected together via several I/O channels. If one of the channels fails, the remaining ones can still be used.
- Instant copies can be used to protect against logical errors. For example, it would be possible to create an instant copy of a database every hour. If a table is 'accidentally' deleted, then the database could revert to the last instant copy in which the database is still complete.
- Remote mirroring protects against physical damage. If, for whatever reason, the original data can no longer be accessed, operation can continue using the data copy that was generated using remote mirroring.

This list shows that disk subsystems can guarantee the availability of data to a very high degree. Despite everything it is in practice sometimes necessary to shut down and switch off a disk subsystem. In such cases, it can be very tiresome to co-ordinate all project groups to a common waiting window, especially if these are distributed over different time zones.

When choosing a RAID type, it is imperative to consider the impact to disk performance and application IOPS.

In both mirrored and parity RAID configurations, every write operation translates into more I/O overhead for the disks which is referred to as *write penalty.* In a RAID 1 implementation, every write operation must be performed on two disks configures as a mirrored pair while in a RAID 5 implementation, a write operation may manifest as four I/O operations. When performing small I/Os to a disk configured with RAID 5, the controller has to read, calculate, and write a parity segment for every data write operation.

Figure 3.24, illustrates a single write operation on RAID 5 that contains a group of five disks. Four of these disks are used for data and one is used for parity.

The parity (*P*) at the controller is calculated as follows:

$$E_P = E_1 + E_2 + E_3 + E_4 \text{ (XOR operations)}$$

Here, D_1 to D_4 is striped data across the RAID group of five disks.

Whenever the controller performs a write I/O, parity must be computed by reading the old parity (E_P old) and the old data (E_4 old) from the disk, which means two read I/Os. The new parity ($E_{P\ new}$) is computed as follows:

$$E_{P\ new} = E_{P\ old} - E_{4\ old} + E_{4\ new} \text{ (XOR operations)}$$

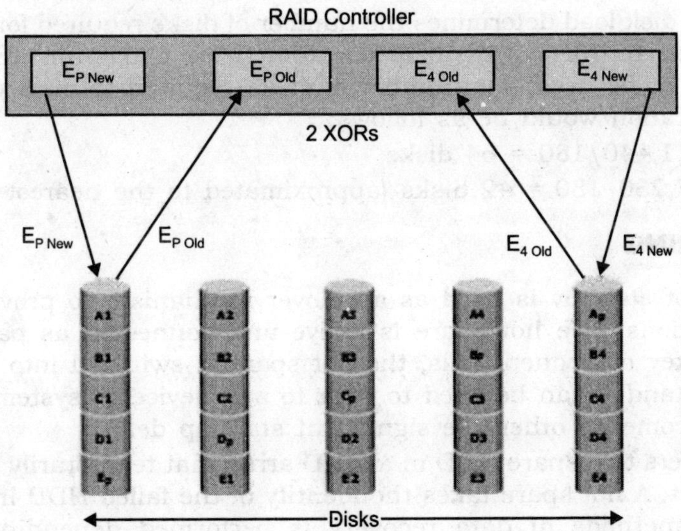

Fig. 3.24.

After computing the new parity, the controller completes the write I/O by writing the new data and the new parity onto the disks, amounting to two write I/Os. Therefore, the controller performs two disk reads and two disk writes for every write operation, and the write parity in RAID 5 implementations is 4.

In RAID 6, which maintains dual parity, a disk write requires three read operations for $E_{P1\ old}$, $E_{P2\ old}$ and $E_{4\ old}$. After calculating $E_{P1\ New}$ and E_{P2} New, the controller performs three write I/O operations for $E_{P1\ New}$, $E_{P2\ New}$ and $E_{4\ New}$. Therefore, in a RAID 6 implementation, the controller performs six I/O operations for each write I/O, and the write penalty is 6.

When deciding the number of disks required for an application, it is important to consider the impact of RAID based on IOPS generated by the application. The totalk disk load should be computed by considering the type of RAID configuration and the ratio of read compared to write from the host.

The following example illustrates the method of computing the disk load in different types of RAID.

Consider an application that generates 5,200 IOPS, with 60 percent of them being reads.

The disk load in RAID 5 is calculated as follows:

RAOD 5 disk load = 0.6 × 5,200 + 4 × (0.4 × 5,200) [because the write penalty for RAID 5 is 4].

= 3,120 + 4 × 2,080
= 3,120 + 8,320
= 11,440 IOPS

The disk load in RAID 1 is calculated as follows:

RAID 1 disk load = 0.6 × 5,200 + 2 × (0.4 × 5,200) [because every write manifests as two writes to the disks]

= 3,120 + 2 × 2,080
= 3,120 + 4,160
= 7,280 IOPS

The computed disk load determines the number of disks required for the application. If in this example an HDD with a specification of a maximum 180 IOPS for the application needs to be used, the number of disks required to meet the workload for the RAID configuration would be as follows:

- RAID 5 : 11,440/180 = 64 disks
- RAID 1 : 7,280/180 = 42 disks (approximated to the nearest even number)

3.9. HOT SPARING

A hot spare or hot standby is used as a failover mechanism to provide reliability in system configurations. The hot spare is active and connected as part of a working system. When a key component fails, the hot spare is switched into operation. More generally, a hot standby can be used to refer to any device or system that is held in readiness to overcome an otherwise significant start-up delay.

A *hot spare* refers to a spare HDD in a RAID array that temporarily replaces a failed HDD of a RAID set. A hot spare takes the identity of the failed HDD in the array. One of the following methods of data recovery is performed depending on the RAID implementation:

- If parity RAID is used, then the data is rebuilt onto the hot spare from the parity and the data on the surviving HDDs in the RAID set.
- If mirroring is used, then the data from the surviving mirror is used to copy the data.

When the failed HDD is replaced with a new HDD, one of the following takes place:

- The hot spare replaces the new HDD permanently. This means that it is no longer a hot spare, and a new hot spare must be configured on the array.
- When a new HDD is added to the system, data from the hot spare is copied to it. The hot spare returns to its idle state, ready to replace the next failed drive.

A hot spare should be large enough to accommodat6e data from a failed drive. Some systems implement multiple hot spares to improve data availability.

A hot spare can be configured as *automatic* or *user initiated*, which specifies how it will be used in the event of disk failure. In an automatic configuration, when the recoverable error rates for a disk exceed a predetermined threshold, the disk subsystem tries to copy data from the failing disk to the hot spare automatically. If this task is completed before the damaged disk fails, then the subsystem switches to the hot spare and \marks the failing disk as unusable. Otherwise, it uses parity or the mirrored disk to recover the data. In the case of a user-initiated configuration, the administrator has control of the rebuild process. For example, the rebuild could occur overnight to prevent any degradation of system performance. However, the system is vulnerable to another failure if a hot spare is unavailable.

SUMMARY

1. Hard disks and tapes are currently the most important media for the storage of data. When storage networks are introduced, the existing small storage devices are replaced by a few large storage systems (storage consolidation). For example, individual hard disks and small disk stacks are replaced by large disk subsystems that can store between a few hundred gigabytes and several ten terabytes of data, depending upon size.
2. The controller of the disk subsystem must ultimately store all data on physical hard disks. Standard hard disks that range in size from 18 GB to 250 GB are currently (2003) used for this purpose. Since the maximum number hard disks that can be used is often limited, the size of the hard disk used gives an indication of the maximum capacity of the overall disk subsystem.
3. Standard I/O techniques such as SCSI and Fibre Channel, to an increasing degree SATA (Serial ATA) and sometimes also SSA (Serial Storage Architecture) are very often used for the internal I/O channels between connection ports and controller and between controller and internal hard disks. Sometimes, however, proprietary – i.e. manufacturer-speci?c – I/O technologies are used.
4. A disk subsystem with a RAID controller offers greater functional scope than a JBOD disk subsystem. RAID was originally developed at a time when hard disks were still very expensive and less reliable than they are today. RAID was originally called 'Redundant Array of Inexpensive Disks'. Today RAID stands for 'Redundant Array of Independent Disks'. Disk subsystems that support RAID are sometimes also called RAID arrays.
5. RAID has two main goals: to increase performance by striping and to increase fault-tolerance by redundancy. Striping distributes the data over several hard disks and thus distributes the load over more hardware.
6. The bundle of physical hard disks brought together by the RAID controller are also known as virtual hard disks. A server that is connected to a RAID system sees only the virtual hard disk;
7. One factor common to almost all RAID levels is that they store redundant information. If a physical hard disk fails, its data can be reconstructed from the hard disks that remain intact. The defective hard disk can even be replaced by a new one during operation if a disk subsystem has the appropriate hardware.
8. In contrast to RAID 0, in RAID 1 fault-tolerance is of primary importance. The basic form of RAID 1 brings together two physical hard disks to form a virtual hard disk by mirroring the data on the two physical hard disks.
9. RAID 10 provides excellent performance at a high level of fault-tolerance. The problem with this is that mirroring using RAID 1 means that all data is written to the physical hard disk twice. RAID 10 thus doubles the required storage capacity.
10. When processing write commands for RAID 4 and RAID 5 arrays, RAID controllers use the above-mentioned mathematical properties of the XOR operation for the recalculation of the parity block.

QUESTION WITH ANSWER

Q.1. Component architecture of intelligent disk subsystems.

Ans. In contrast to a file server, a disk subsystem can be visualized as a hard disk server. Servers are connected to the connection port of the disk subsystem using standard I/O techniques such as SCSI, Fibre Channel or iSCSI and can thus use the storage capacity that the disk subsystem provides Fig. 3.25. The internal structure of the disk subsystem is completely hidden from the server, which sees only the hard disks that the disk subsystem provides to the server.

The connection ports are extended to the hard disks of the disk subsystem by means of internal I/O channels Fig. 3.26. In most disk subsystems there is a controller between the connection ports and the hard disks. The controller can significantly increase the data availability and data access performance with the aid of a so-called RAID procedure (RAID = Redundant Array of Independent Disks). Furthermore, some controllers realize the copying services instant copy and remote mirroring and further additional services. The controller uses a cache in an attempt to accelerate read and write accesses to the server.

All sizes of disk subsystems are available. Small disk subsystems have one to two connections for servers or storage networks, six to eight hard disks are depending.

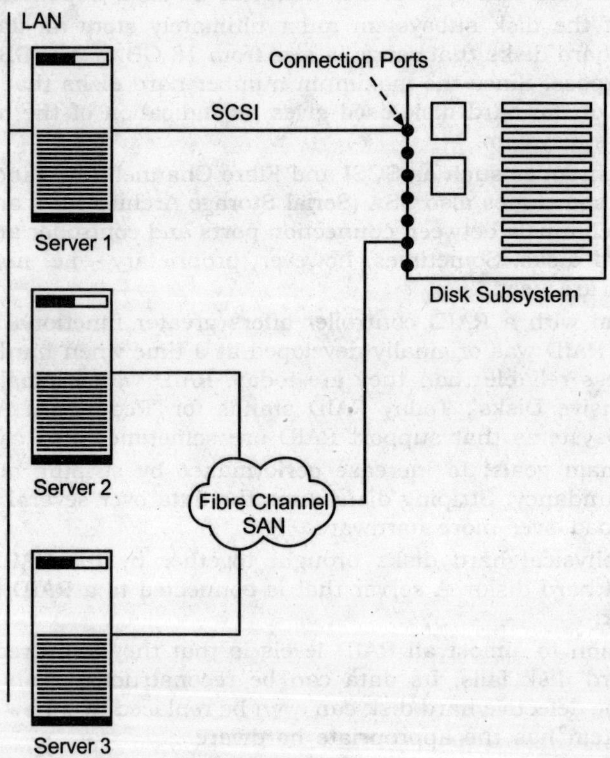

Fig. 3.25.

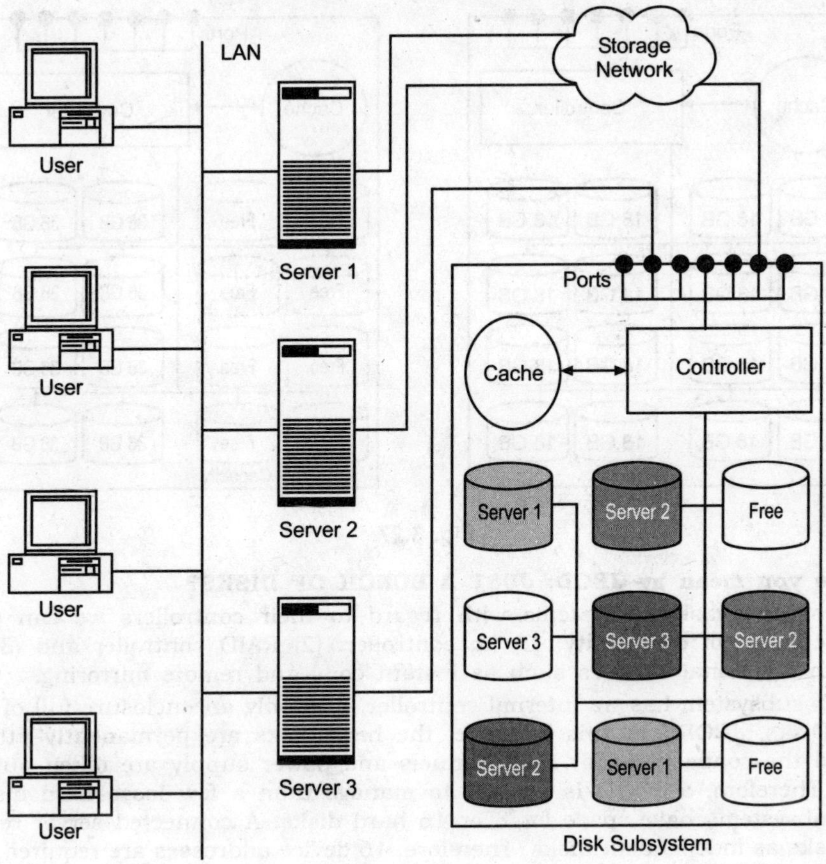

Fig. 3.26.

Q.2. Explain Disk physical structure components.

Ans. The controller of the disk subsystem must ultimately store all data on physical hard disks.

Standard hard disks that range in size from 18 GB to 250 GB are currently (2003) used for this purpose. Since the maximum number hard disks that can be used is often limited, the size of the hard disk used gives an indication of the maximum capacity of the overall disk subsystem.

When selecting the size of the internal physical hard disks it is necessary to weigh the requirements of maximum performance against those of the maximum capacity of the overall system. With regard to performance it is often beneficial to use smaller hard disks at the expense of the maximum capacity: given the same capacity, if more hard disks are available in a disk subsystem, the data is distributed over several hard disks and thus the overall load is spread over more arms and read/write heads and usually over more I/O channels (Figure 3.27). For most applications, medium-sized hard disks are sufficient. Only for applications with extremely high performance requirements should smaller hard disks.

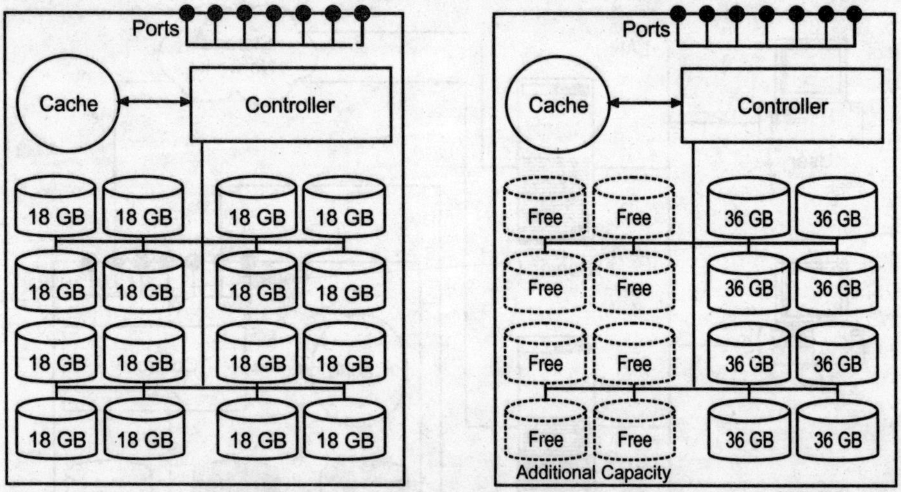

Fig. 3.27.

Q.3. What do you mean by JBOD: JUST A BUNCH OF DISKS?

Ans. If we compare disk subsystems with regard to their controllers we can differentiate between three levels of complexity: (1) no controller; (2) RAID controller and (3) intelligent controller with additional services such as instant copy and remote mirroring.

If the disk subsystem has no internal controller, it is only an enclosure full of disks (Just a Bunch of Disks, JBOD). In this instance, the hard disks are permanently fitted into the enclosure and the connections for I/O channels and power supply are taken outwards at a single point. Therefore, a JBOD is simpler to manage than a few loose hard disks. Typical JBOD disk subsystems have space for 8 or 16 hard disks. A connected server recognizes all these hard disks as independent disks. Therefore, 16 device addresses are required for a JBOD disk subsystem incorporating 16 hard disks. In some I/O techniques such as SCSI and Fibre Channel arbitrated loop, this can lead to a bottleneck at device addresses.

In contrast to intelligent disk subsystems, a JBOD disk subsystem in particular is not capable of supporting RAID or other forms of virtualization. If required, however, these can be realized outside the JBOD disk subsystem, for example, as software in the server or as an independent virtualization entity in the storage network.

Q.4. Define Raid Levels and Parity Algorithms.

Ans. A disk subsystem with a RAID controller offers greater functional scope than a JBOD disk subsystem. RAID was originally developed at a time when hard disks were still very expensive and less reliable than they are today. RAID was originally called 'Redundant Array of Inexpensive Disks'. Today RAID stands for 'Redundant Array of Independent Disks'. Disk subsystems that support RAID are sometimes also called RAID arrays.

RAID has two main goals: to increase performance by striping and to increase fault-tolerance by redundancy. Striping distributes the data over several hard disks and thus distributes the load over more hardware. Redundancy means that additional information is stored so that the operation of the application itself can continue in the event of the failure of a hard disk. You cannot increase the performance of an individual hard disk any more than you can improve its fault-tolerance. Individual physical hard disks are slow and have a limited life-cycle. However, through a suitable combination of physical hard disks it is possible to significantly increase the fault-tolerance and performance of the system as a whole.

The bundle of physical hard disks brought together by the RAID controller are also known as virtual hard disks. A server that is connected to a RAID system sees only the virtual hard disk; the fact that the RAID controller actually distributes the data over several physical hard disks is completely hidden to the server Fig. 3.28. This is only visible to the administrator from outside.

A RAID controller can distribute the data that a server writes to the virtual hard disk amongst the individual physical hard disks in various manners. These different procedures are known as RAID levels. explains various RAID levels in detail.

One factor common to almost all RAID levels is that they store redundant information. If a physical hard disk fails, its data can be reconstructed from the hard disks that remain intact. The defective hard disk can even be replaced by a new one during operation if a disk subsystem has the appropriate hardware. Then the RAID controller reconstructs the data of the exchanged hard disk. This process remains hidden to the server apart from a possible reduction in performance: the server can continue to work uninterrupted on the virtual hard disk.

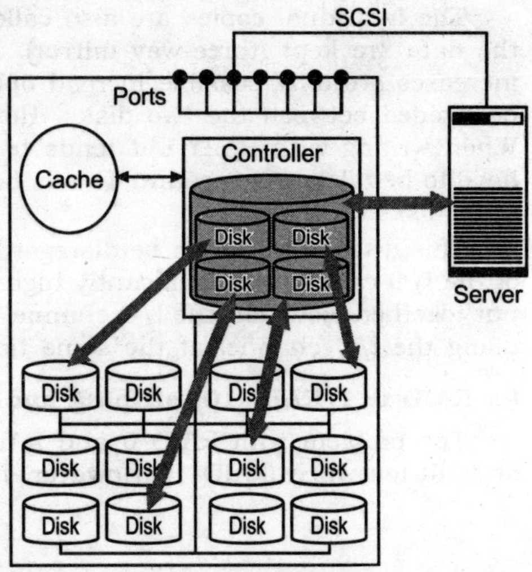

Fig. 3.28.

Modern RAID controllers initiate this process automatically. This requires the definition of so-called hot spare disks Fig. 3.28. The hot spare disks are not used in normal operation. If a disk fails, the RAID controller immediately begins to copy the data of the remaining intact disk onto a hot spare disk. After the replacement of the defective disk, this is included in the pool of hot spare disks. Modern RAID controllers can manage a.

Q.5. Define Different Raid Levels in Detail.

Ans. RAID has developed since its original definition in 1987. Due to technical progress some RAID levels are now practically meaningless, whilst others have been modified or added at a later date. This section introduces the RAID levels that are currently the most significant in practice. We will not introduce RAID levels that represent manufacturer specific variants and variants that only deviate slightly from the basic forms mentioned in the following.

(*a*) RAID 0: block-by-block striping

RAID 0 distributes the data that the server writes to the virtual hard disk onto one physical hard disk after another block-by-block (block-by-block striping). Figure 3.9 shows a RAID array with four physical hard disks. In Figure 3.9 the server writes the blocks A, B, C, D, E, etc. onto the virtual hard disk one after the other. The RAID controller distributes the sequence of blocks onto the individual physical hard disks: it writes the first block, A, to the first physical hard disk, the second block, B, to the second physical hard disk, block C to the third and block D to the fourth. Then it begins to write to the first physical hard disk once again, writing block E to the first disk, block F to the second, and so on.

RAID 0 increases the performance of the virtual hard disk as follows: the individual hard disks can exchange data with the RAID controller via the I/O channel significantly more quickly than they can write to or read from the rotating disk. In Figure 3.9 the RAID controller sends the first block, block A, to the first hard disk. This takes some time to write the block to the disk. Whilst the first disk is writing the first block to the physical hard disk, the RAID controller is already sending the second block, block B,

(*b*) RAID 1: block-by-block mirroring

In contrast to RAID 0, in RAID 1 fault-tolerance is of primary importance. The basic form of RAID 1 brings together two physical hard disks to form a virtual hard disk by mirroring the data on the two physical hard disks. If the server writes a block to the virtual hard disk, the RAID controller writes this block to both physical hard disks Fig. 3.10.

The individual copies are also called mirrors. Normally, two or sometimes three copies of the data are kept (three-way mirror). In a normal operation with pure RAID 1, performance increases are only possible in read operations. After all, when reading the data the load can be divided between the two disks. However, this gain is very low in comparison to RAID 0. When writing with RAID 1 it tends to be the case that reductions in performance may even have to be takn into account. This is because the RAID controller has to send the data to both hard disks.

This disadvantage can be disregarded for an individual write operation, since the capacity of the I/O channel is significantly higher than the maximum write speed of the two hard disks put together. However, the I/O channel is under twice the load, which hinders other data traffic using the I/O channel at the same time.

(c) RAID 0+1/RAID 10: striping and mirroring combined

The problem with RAID 0 and RAID 1 is that they increase either performance (RAID 0) or fault-tolerance (RAID 1). However, it would be nice to have both performance and

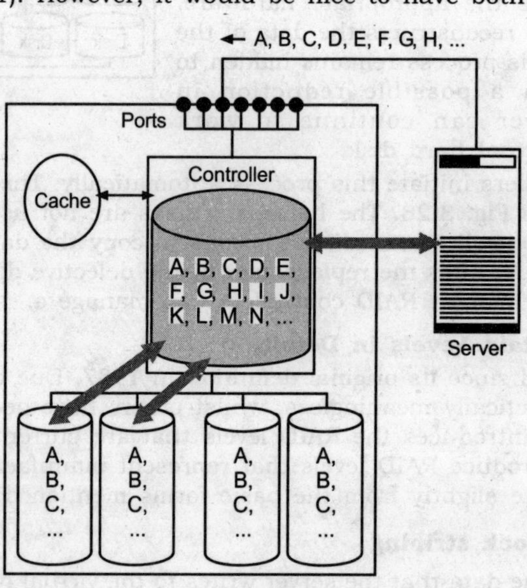

Fig. 3.29.

(d) RAID 4 and RAID 5: parity instead of mirroring

RAID 10 provides excellent performance at a high level of fault-tolerance. The problem with this is that mirroring using RAID 1 means that all data is written to the physical hard disk twice. RAID 10 thus doubles the required storage capacity.

The idea of RAID 4 and RAID 5 is to replace all mirror disks of RAID 10 with a single parity hard disk. Figure 3.30 shows the principle of RAID 4 based upon five physical hard disks. The server again writes the blocks A, B, C, D, E, etc. to the virtual hard disk sequentially. The RAID controller stripes the data blocks over the first four physical hard disks. Instead of mirroring all data onto the further four physical hard disks, as in RAID 10, the RAID controller calculates a parity block for every four blocks and writes this onto the fifth physical hard disk. For example, the RAID controller calculates the parity block PABCDfor the blocks A, B, C and D. If one of the four data disks fails, the RAID controller can reconstruct the data of the defective disks using the three other data disks and the parity disk. In comparison to the examples in Figures 3.11 (RAID 0+1) and 3.12 (RAID 10), RAID 4 saves three physical hard disks. As in all other RAID levels, the server again sees only the virtual disk, as if it were a single physical hard disk.

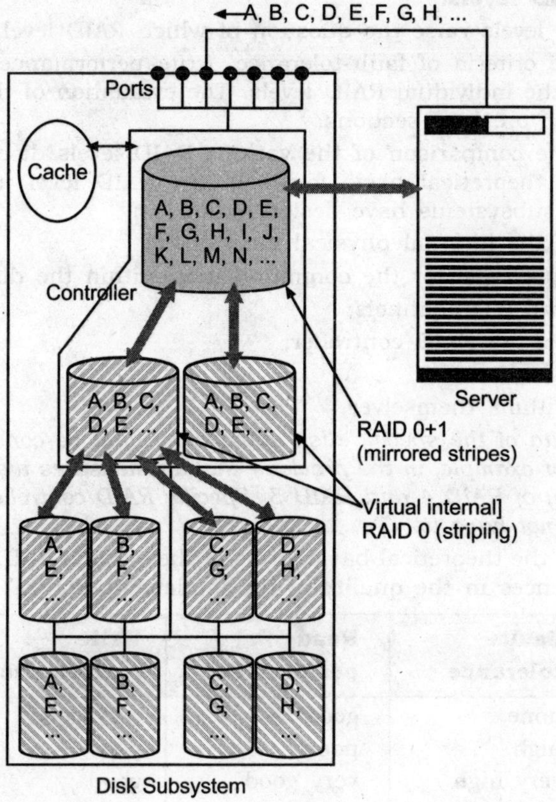

Fig. 3.30.

Q.6. Compare RAID 2 and RAID 3.

Ans. When introducing the RAID levels we are sometimes asked: 'and what about RAID 2 and RAID 3?'. The early work on RAID began at a time when disks were not yet very reliable: bit errors were possible that could lead to a written 'one' being read as 'zero' or a written 'zero' being read as 'one'. In RAID 2 the Hamming code is used, so that redundant information is stored in addition to the actual data. This additional data permits the recognition of read errors and to some degree also makes it possible to correct them. Today, comparable functions are performed by the controller of each individual hard disk, which means that RAID 2 no longer has any practical significance.

Like RAID 4 or RAID 5, RAID 3 stores parity data. RAID 3 distributes the data of a block amongst all the disks of the RAID 3 system so that, in contrast to RAID 4 or RAID 5, all disks are involved in every read or write access. RAID 3 only permits the reading and writing of whole blocks, thus dispensing with the write penalty that occurs in RAID.

4 and RAID 5. The writing of individual blocks of a parity group is thus not possible. In addition, in RAID 3 the rotation of the individual hard disks is synchronized so that the data of a block can truly be written simultaneously. RAID 3 was for a long time called the recommended RAID level for sequential write and read profiles such as data mining and video processing. Current hard disks come with a large cache of their own, which means that they can temporarily store the data of an entire track, and they have significantly higher rotation speeds than the hard disks of the past. As a result of these innovations, other RAID levels are now suitable for sequential load profiles, meaning that RAID 3 is becoming less and less important.

Q.7. Compare the RAID levels.

Ans. The various RAID levels raise the question of which RAID level should be used when.

Table compares the criteria of fault-tolerance, write performance, read performance and space requirement for the individual RAID levels. The evaluation of the criteria can be found in the discussion in the previous sections.

Caution Please: The comparison of the various RAID levels discussed in this section is only applicable to the theoretical basic forms of the RAID level in question. in practice, manufacturers of disk subsystems have design options in

- the selection of the internal physical hard disks;
- the I/O technique used for the communication within the disk subsystem;
- the use of several I/O channels;
- the realization of the RAID controller;
- the size of the cache; and
- the cache algorithms themselves.

The performance data of the specific disk subsystem must be considered very carefully for each individual case. For example, in the previous chapter measures were discussed that greatly reduce the write penalty of RAID 4 and RAID 5. Specific RAID controllers may implement these measures, but they do not have to.

The table compares the theoretical basic forms of the various RAID levels. In practice there are very marked differences in the quality of the implementation of RAID controllers

RAID level	Fault tolerance	Read performance	Write performance	Space requirement
RAID 0	none	good	very good	minimal
RAID 1	high	poor	poor	high
RAID 10	very high	very good	good	high
RAID 4	high	good	very very poor	low
RAID 5	high	good	very poor	low

Subject to the above warning, RAID 0 is the choice for applications for which the maximum write performance is more important than protection against the failure of a disk. Examples are the storage of multimedia data for film and video production and the recording of physical experiments in which the entire series of measurements has no value if all measured values cannot be recorded. In this case it is more beneficial to record all of the measured data on a RAID 0 array first and then copy it after the experiment, for example on a RAID 5 array. In databases, RAID 0 is used as a fast store for segments in which intermediate results for complex requests are to be temporarily stored. However, as a rule hard disks tend to fail at the most inconvenient moment so database administrators only use RAID 0 if it is absolutely necessary, even for temporary data.

With RAID 1, performance and capacity are limited because only two physical hard disks are used. RAID 1 is therefore a good choice for small databases for which the configuration of a virtual RAID 5 or RAID 10 disk would be too large. A further important field of application for RAID 1 is in combination with RAID 0.

RAID 10 is used in situations where high write performance and high fault-tolerance are called for. For a long time it was recommended that database log files be stored on RAID 10. Databases record all changes in log files so this application has a high write component. After a system crash the restarting of the database can only be guaranteed if all log files are fully available. Manufacturers of storage systems disagree as to whether this recommendation is still valid as there are now fast RAID 4 and RAID 5 implementations.

RAID 4 and RAID 5 save disk space at the expense of a poorer write performance. For a long time the rule of thumb was to use RAID 5 where the ratio of read operations to write operations is 70 : 30. At this point we wish to repeat that there are now storage systems on the market with excellent write performance that store the data internally using RAID 4 or RAID 5.

OBJECTIVE TYPE QUESTIONS

1. How can a CIFS share be hidden from network browsing?
   (a) Append the share name with a "$"
   (b) Edit the permissions of the share and select "Deny for Browsing"
   (c) Edit the share's properties and select "Hidden"
   (d) Prefix the share name with a "."

2. A NAS solution that includes high availability and network connections that are always available is required. The Ethernet switch is made by Typhoon Data Networks and is IEEE channel compliant.
   Which EMC. NAS networking solution must be implemented?
   (a) FastEtherChannel (b) HSRP
   (c) LACP (d) VLAN

3 Which two objects are associated together to provide name resolution?\
   (a) Hostname and IP address (b) B. Hostname and MAC. address
   (c) Hostname and network port
   (d) Hostname and operating system version number

4. Which state should a Virtual Data Mover be in before it is fully functional?
   (a) Loaded (b) Mounted
   (c) PermUnloaded (d) TempUnloade

5. A UNIX client mounts a file system presented by the Celerra. Which name resolution is valid?
   (a) CIFS (b) NFS
   (c) NIS (d) WINS

ANSWERS

1. (a) 2. (c) 3. (a) 4. (a) 5. (c)

CHAPTER

4

INTRODUCTION TO NETWORKED STORAGE

Inside this Chapter

- Introduction
- JBOD: Just A Bunch of Disks
- Direct-Attached Storage (DAS), its Evolution and Comparison
- Network-attached storage (NAS), its Evolution and Comparison
- Introduction to Networked Storage (SAN)
- Content-Addressed Storage, its Evolution and Comparison
- Standards, Management, Security and Limitations of DAS, NAS, CAS and SAN

4.0. INTRODUCTION

Computing is based on information. Information is the underlying resource on which all computing processes are based; it is a company asset. Information is stored on storage media, and is accessed by applications executing on a server. Often the information is a unique company asset. Information is created and acquired every second of every day. Information is the currency of business. To ensure that any business delivers the expected results, they must have access to accurate information, and without delay. The management and protection of business information is vital for the availability of business processes. This chapter introduces the concept of a Storage Area Network, which has been regarded as the ultimate response to all these needs.

4.1. JBOD: JUST A BUNCH OF DISKS

If we compare disk subsystems with regard to their controllers we can differentiate between three levels of complexity: (1) no controller; (2) RAID controller and (3) intelligent controller with additional services such as instant copy and remote mirroring.

If the disk subsystem has no internal controller, it is only an enclosure full of disks (Just a Bunch of Disks, JBOD). In this instance, the hard disks are permanently fitted into the enclosure and the connections for I/O channels and power supply are taken outwards at a single point. Therefore, a JBOD is simpler to manage than a few loose hard disks. Typical JBOD disk subsystems have space for 8 or 16 hard disks. A connected server recognizes all these hard disks as independent disks. Therefore, 16 device addresses are required for a JBOD disk subsystem incorporating 16 hard disks. In some I/O techniques such as SCSI and Fibre Channel arbitrated loop, this can lead to a bottleneck at device addresses.

In contrast to intelligent disk subsystems, a JBOD disk subsystem in particular is not capable of supporting RAID or other forms of virtualization. If required, however, these can be realized outside the JBOD disk subsystem, for example, as software in the server or as an independent virtualization entity in the storage network.

4.2. DIRECT-ATTACHED STORAGE (DAS) , ITS EVOLUTION AND COMPARISON

Direct-Attached Storage (DAS is an architecture where storage connects directly to servers. Applications access data from DAS using block-level access protocols. The internal HDD of a host, tape libraries, and directly connected external HDD packs are some examples of DAS.

Although the implementation of storage networking technologies are gaining popularity, DAS has remained ideal for localized data access and sharing in environments that have a small number of servers. For example, small businesses, departments, and workgroups that do not share information across enterprises find DAS to be an appropriate solution. Medium-size companies use DAS for file serving and e-mail, while larger enterprises leverage DAS in conjunction with SAN and NAS.

This chapter details the two types of DAS along with their benefits and limitations. A major part of this chapter is devoted to detailing the types of storage devices used in DAS environments and SCSI—the most prominent protocol used in DAS.

4.2.1. DAS and its Application, Components

DAS is classified as internal or external, based on the location of the storage device with respect to the host.

4.2.1.1. Internal DAS

In *internal DAS* architectures, the storage device is internally connected to the host by a serial or parallel bus. The physical bus has distance limitations and can only be sustained over a shorter distance for high-speed connectivity. In addition, most internal buses can support only a limited number of devices, and they occupy a large amount of space inside the host, making maintenance of other components difficult.

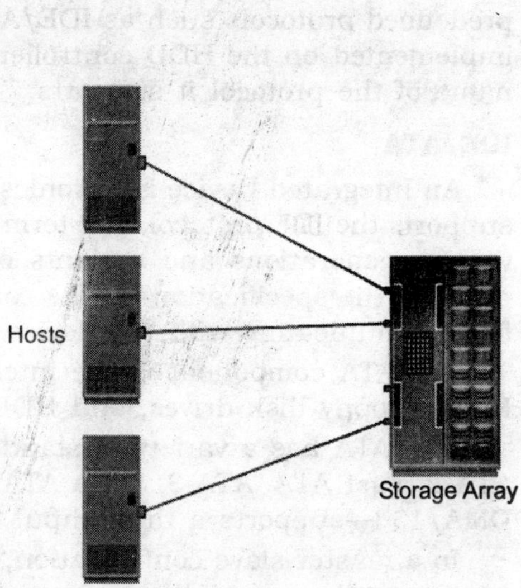

Fig. 4.1.

4.2.1.2. External DAS

In *external DAS* architectures, the server connects directly to the external storage device (see Fig. 4.1.). In most cases, communication between the host and the storage device takes place over SCSI or FC protocol. Compared to internal DAS, an external DAS overcomes the distance and device count limitations and provides centralized management of storage devices.

4.2.2. DAS Security and Limitations

DAS requires a relatively lower initial investment than storage networking. Storage networking architectures are discussed later in this book. DAS configuration is simple and can be deployed easily and rapidly. Setup is managed using host-based tools, such as the host OS, which makes storage management tasks easy for small and medium enterprises. DAS is the simplest solution when compared to other storage networking models and requires fewer management tasks, and less hardware and software elements to set up and operate.

However, DAS does not scale well. A storage device has a limited number of ports, which restricts the number of hosts that can directly connect to the storage. A limited bandwidth in DAS restricts the available I/O processing capability. When capacities are being reached, the service availability may be compromised, and this has a ripple effect on the performance of all hosts attached to that specific device or array. The distance limitations associated with implementing DAS because of direct connectivity requirements can be addressed by using Fibre Channel connectivity. DAS does not make optimal use of resources due to its limited ability to share front end ports. In DAS environments, unused resources cannot be easily re-allocated, resulting in islands of over-utilized and under-utilized storage pools.

Disk utilization, throughput, and cache memory of a storage device, along with virtual memory of a host govern the performance of DAS. RAID-level configurations, storage controller protocols, and the efficiency of the bus are additional factors that affect the performance of DAS. The absence of storage interconnects and network latency provide DAS with the potential to outperform other storage networking configurations.

4.2.3. DAS Elements, Connectivity, Standards & Management

The host and the storage device in DAS communicate with each other by using predefined protocols such as IDE/ATA, SATA, SAS, SCSI, and FC. These protocols are implemented on the HDD controller. Therefore, a storage device is also known by the name of the protocol it supports.

IDE/ATA

An Integrated Device Electronics/Advanced Technology Attachment (IDE/ATA) disk supports the IDE protocol. The term IDE/ATA conveys the dual-naming conventions for various generations and variants of this interface. The IDE component in IDE/ATA provides the specification for the controllers connected to the computer's motherboard for communicating with the device attached.

The ATA component is the interface for connecting storage devices, such as CD-ROMs, floppy disk drives, and HDDs, to the motherboard.

IDE/ATA has a variety of standards and names, such as ATA, ATA/ATAPI, EIDE, ATA-2, Fast ATA, ATA-3, Ultra ATA, and Ultra DMA. The latest version of ATA—Ultra DMA/133—supports a throughput of 133 MB per second.

In a master-slave configuration, an ATA interface supports two storage devices per connector. However, if the performance of the drive is important, sharing a port between two devices is not recommended.

SATA

A SATA (Serial ATA) is a serial version of the IDE/ATA specification. SATA is a disk-interface technology that was developed by a group of the industry's leading vendors with the aim of replacing parallel ATA.

A SATA provides point-to-point connectivity up to a distance of one meter and enables data transfer at a speed of 150 MB/s. Enhancements to the SATA have increased the data transfer speed up to 600 MB/s.

A SATA bus directly connects each storage device to the host through a dedicated link, making use of *low-voltage differential signaling (LVDS).* LVDS is an electrical signaling system that can provide high-speed connectivity over low-cost, twisted-pair copper cables. For data transfer, a SATA bus uses LVDS with a voltage of 250 mV.

A SATA bus uses a small 7-pin connector and a thin cable for connectivity.A SATA port uses 4 signal pins, which improves its pin efficiency compared to the parallel ATA that uses 26 signal pins, for connecting an 80-conductor ribbon cable to a 40-pin header connector.

SATA devices are *hot-pluggable,* which means that they can be connected or removed while the host is up and running. A SATA port permits single-device connectivity. Connecting multiple SATA drives to a host requires multiple ports to be present on the host. Single-device connectivity enforced in SATA, eliminates the performance problems caused by cable or port sharing in IDE/ATA.

Parallel SCSI

SCSI is available in a variety of interfaces. Parallel SCSI (referred to as SCSI) is one of the oldest and most popular forms of storage interface used in hosts. SCSI is a set of standards used for connecting a peripheral device to a computer and transferring data between them. Often, SCSI is used to connect HDDs and tapes to a host. SCSI can also connect a wide variety of other devices such as scanners and printers. Communication between the hosts and the storage devices uses the SCSI command set, described later in this chapter.

Since its inception, SCSI has undergone rapid revisions, resulting in continuous performance improvements. The oldest SCSI variant, called SCSI-1 provided data transfer rate of 5 MB/s; SCSI Ultra 320 provides data transfer speeds of 320 MB/s. Other variants of SCSI and transfer speeds are listed in.

Feature	IDE/ATA	SCSI
Speed	100, 133, 150 MB/s	320 MB/s
Connectivity	Internal	Internal and external
Cost	Low	Moderate to high
Hot-pluggable	No	Yes
Performance	Moderate to low	High
Ease of configuration	High	Low to moderate
Maximum number of devices suported	2	16

4.3. NETWORK-ATTACHED STORAGE (NAS), ITS EVOLUTION AND COMPARISON

Network-attached storage (NAS) is an IP-based file-sharing device attached to a local area network. NAS provides the advantages of server consolidation by eliminating the need for multiple file servers. It provides storage consolidation through file-level data access and sharing. NAS is a preferred storage solution that enables clients to share files quickly and directly with minimum storage management overhead. NAS also helps to eliminate bottlenecks that users face when accessing files from a general-purpose server.

NAS uses network and file-sharing protocols to perform filing and storage functions. These protocols include TCP/IP for data transfer and CIFS and NFS for remote file service. NAS enables both UNIX and Microsoft Windows users to share the same data seamlessly. To enable data sharing, NAS typically uses NFS for UNIX, CIFS for Windows, and File Transfer Protocol (FTP) and other protocols for both environments. Recent advancements in networking technology have enabled NAS to scale up to enterprise requirements for improved performance and reliability in accessing data.

A NAS device is a dedicated, high-performance, high-speed, single-purpose file serving and storage system. NAS serves a mix of clients and servers over an IP network. Most NAS devices support multiple interfaces and networks.

A NAS device uses its own operating system and integrated hardware, software components to meet specific file service needs. Its operating system is optimized for file I/O and, therefore, performs file I/O better than a generalpurpose server. As a result, a NAS device can serve more clients than traditional file servers, providing the benefit of server consolidation.

4.3.1. NAS and its Application, Components

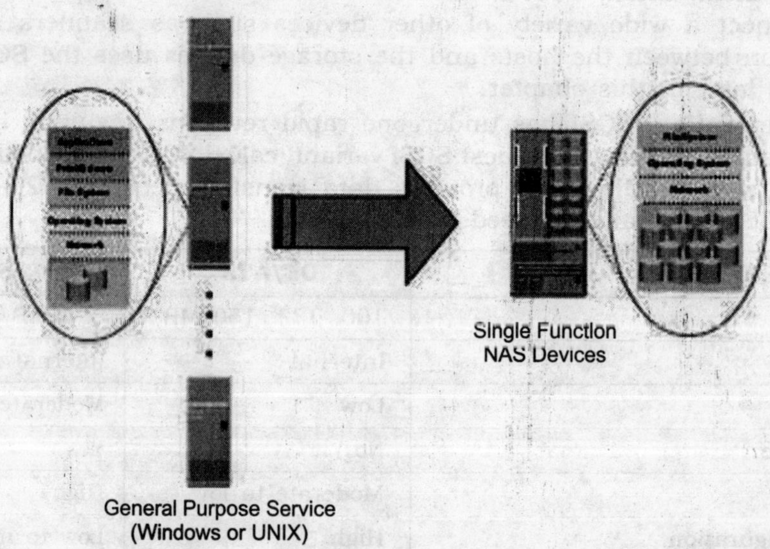

Fig. 4.2.

A NAS device is optimized for file-serving functions such as storing, retrieving, and accessing files for applications and clients. As shown in Fig. 4.2, a general-purpose server can be used to host any application, as it runs a generic operating system. Unlike a general-purpose server, a NAS device is dedicated to file-serving. It has a real-time operating system dedicated to file serving by using open-standard protocols. Some NAS vendors support features such as native clustering for high availability.

4.3.2. NAS Security and Limitations

NAS offers the following benefits:

- **Supports comprehensive access to information:** Enables efficient file sharing and supports many-to-one and one-to-many configurations. The many-to-one configuration enables a NAS device to serve many clients simultaneously. The one-to-many configuration enables one client to connect with many NAS devices simultaneously.
- **Improved efficiency:** Eliminates bottlenecks that occur during file access from a general-purpose file server because NAS uses an operating system specialized for file serving. It improves the utilization of general-purpose servers by relieving them of file-server operations.
- **Improved flexibility:** Compatible for clients on both UNIX and Windows platforms using industry-standard protocols. NAS is flexible and can serve requests from different types of clients from the same source.
- **Centralized storage:** Centralizes data storage to minimize data duplication on client workstations, simplify data management, and ensures greater data protection.
- **Simplified management:** Provides a centralized console that makes it possible to manage file systems efficiently.
- **Scalability:** Scales well in accordance with different utilization profiles and types of business applications because of the high performance and low-latency design.
- **High availability:** Offers efficient replication and recovery options, enabling high data availability. NAS uses redundant networking components that provide maximum connectivity options. A NAS device can use clustering technology for failover.
- **Security:** Ensures security, user authentication, and file locking in conjunction with industry-standard security schemas.

4.3.3. NAS Elements, Connectivity, Standards & Management

As mentioned earlier, there are two types of NAS implementations: integrated and gateway. The *integrated NAS* device has all of its components and storage system in a single enclosure. In *gateway* implementation, NAS head shares its storage with SAN environment.

Integrated NAS

An integrated NAS device has all the components of NAS, such as the NAS head and storage, in a single enclosure, or frame. This makes the integrated NAS a self-contained environment. The NAS head connects to the IP network to provide connectivity to the clients and service the file I/O requests. The storage consists of a number of disks that can range from low-cost ATA to highthroughput FC disk drives. Management software manages the NAS head and storage configurations.

An integrated NAS solution ranges from a low-end device, which is a single enclosure, to a high-end solution that can have an externally connected storage array.

A low-end appliance-type NAS solution is suitable for applications that a small department may use, where the primary need is consolidation of storage, rather than high performance or advanced features such as disaster recovery and business continuity. This solution is fixed in capacity and might not be upgradable beyond its original configuration. To expand the capacity, the solution must be scaled by deploying additional units, a task that increases management overhead because multiple devices have to be administered.

In a high-end NAS solution, external and dedicated storage can be used. This enables independent scaling of the capacity in terms of NAS heads or storage. However, there is a limit to scalability of this solution.

Gateway NAS

A gateway NAS device consists of an independent NAS head and one or more storage arrays. The NAS head performs the same functions that it does in the integrated solution; while the storage is shared with other applications that require block-level I/O. Management functions in this type of solution are more complex than those in an integrated environment because there are separate administrative tasks for the NAS head and the storage. In addition to the components that are explicitly tied to the NAS solution, a gateway solution can also utilize the FC infrastructure, such as switches, directors, or direct-attached storage arrays.

The gateway NAS is the most scalable because NAS heads and storage arrays can be independently scaled up when required. Adding processing capacity to the NAS gateway is an example of scaling. When the storage limit is reached, it can scale up, adding capacity on the SAN independently of the NAS head.

Administrators can increase performance and I/O processing capabilities for their environments without purchasing additional interconnect devices and storage. Gateway NAS enables high utilization of storage capacity by sharing it with SAN environment.

Integrated NAS Connectivity

An integrated solution is self-contained and can connect into a standard IP network. Although the specifics of how devices are connected within a NAS implementation vary by vendor and model. In some cases, storage is embedded within a NAS device and is connected to the NAS head through internal connections, such as ATA or SCSI controllers. In others, the storage may be external but connected by using SCSI controllers. In a high-end integrated NAS model, external storage can be directly connected by FC HBAs or by dedicated FC switches. In the case of a low-end integrated NAS model, backup traffic is shared on the same public IP network along with the regular client access traffic.

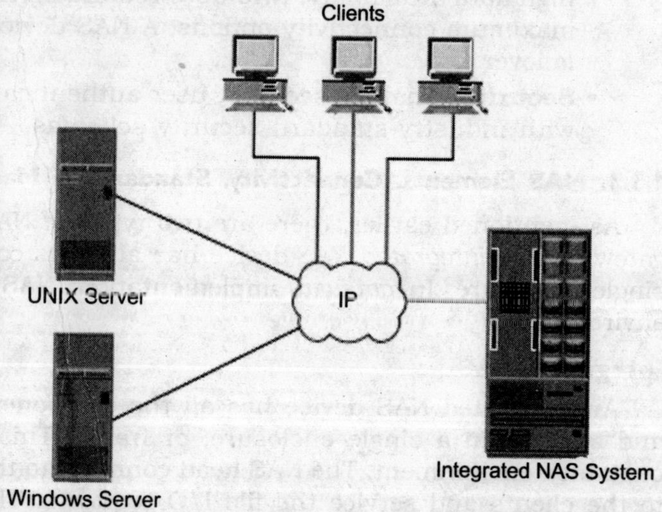

Fig. 4.3.

In the case of a high-end integrated NAS model, an isolated backup network can be used to segment the traffic from impeding client access. More complex solutions may include an intelligent storage subsystem, enabling faster backup and larger capacities while simultaneously enhancing performance.

Gateway NAS Connectivity

In a gateway solution, front-end connectivity is similar to that in an integrated solution. An integrated environment has a fixed number of NAS heads, making it relatively easy to determine IP networking requirements. In contrast, networking requirements in a gateway environment are complex to determine due to scalability options. Adding more NAS heads may require additional networking connectivity and bandwidth.

Communication between the NAS gateway and the storage system in a gateway solution is achieved through a traditional FC SAN. To deploy a stable NAS solution, factors such as multiple paths for data, redundant fabrics, and load distribution must be considered. Figure 4.4, illustrates an example of gateway NAS connectivity.

Implementation of a NAS gateway solution requires analysis of current SAN environment. This analysis is required to determine the feasibility of introducing a NAS workload to the existing SAN. Analyze the SAN to determine whether the workload is primarily read or write, or random or sequential.

Determine the predominant I/O size in use. In general, sequential workloads have large I/Os. Typically, NAS workloads are random with small I/O size. Introducing sequential workload with random workloads can be disruptive to the sequential workload. Therefore, it is recommended to separate the NAS and SAN disks. Also, determine whether the NAS workload performs adequately with the configured cache in the storage subsystem.

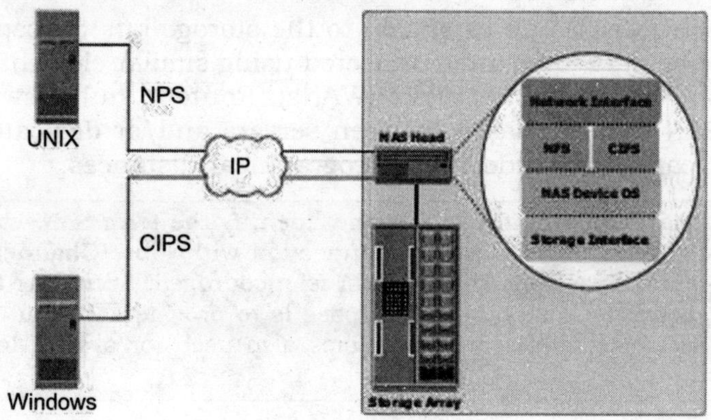

Fig. 4.4.

A NAS device has the following components:

- NAS head (CPU and Memory)
- One or more network interface cards (NICs), which provide connectivity to the network. Examples of NICs include Gigabit Ethernet, Fast Ethernet, ATM, and Fiber Distributed Data Interface (FDDI).
- An optimized operating system for managing NAS functionality
- NFS and CIFS protocols for file sharing

- Industry-standard storage protocols to connect and manage physical disk resources, such as ATA, SCSI, or FC

The NAS environment includes clients accessing a NAS device over an IP network using standard protocols.

4.4. INTRODUCTION TO NETWORKED STORAGE (SAN)

The Storage Network Industry Association (SNIA) defines the SAN as a network whose primary purpose is the transfer of data between computer systems and storage elements. A SAN consists of a communication infrastructure, which provides physical connections; and a management layer, which organizes the connections, storage elements, and computer systems so that data transfer is secure and robust. The term SAN is usually (but not necessarily) identified with block I/O services rather than file access services.

A SAN can also be a storage system consisting of storage elements, storage devices, computer systems, and/or appliances, plus all control software, communicating over a network. Put in simple terms, a SAN is a specialized, high-speed network attaching servers and storage devices and, for this reason, It is sometimes referred to as "the network behind the servers." A SAN allows "any-to-any" connection across the network, using interconnect elements such as routers, gateways, hubs, switches and directors. It eliminates the traditional dedicated connection between a server and storage, and the concept that the server effectively "owns and manages" the storage devices. It also eliminates any restriction to the amount of data that a server can access, currently limited by the number of storage devices attached to the individual server. Instead, a SAN introduces the flexibility of networking to enable one server or many heterogeneous servers to share a common storage utility, which may comprise many storage devices, including disk, tape, and optical storage. Additionally, the storage utility may be located far from the servers that use it.

The SAN can be viewed as an extension to the storage bus concept, which enables storage devices and servers to be interconnected using similar elements as in local area networks (LANs) and wide area networks (WANs): Routers, hubs, switches, directors, and gateways. A SAN can be shared between servers and/or dedicated to one server. It can be local, or can be extended over geographical distances.

> **Note:** The SNIA definition specifically does not identify the term SAN with Fibre Channel technology. When the term SAN is used in connection with Fibre Channel technology, use of a qualified phrase such as *Fibre Channel SAN* is encouraged. According to this definition, an Ethernet-based networkwhose primary purpose is to provide access to storage elements would be considered a SAN. SANs are sometimes also used for system interconnection in clusters.

The diagram in Fig. 4.5. shows a tiered overview of a SAN connecting multiple servers to multiple storage systems.

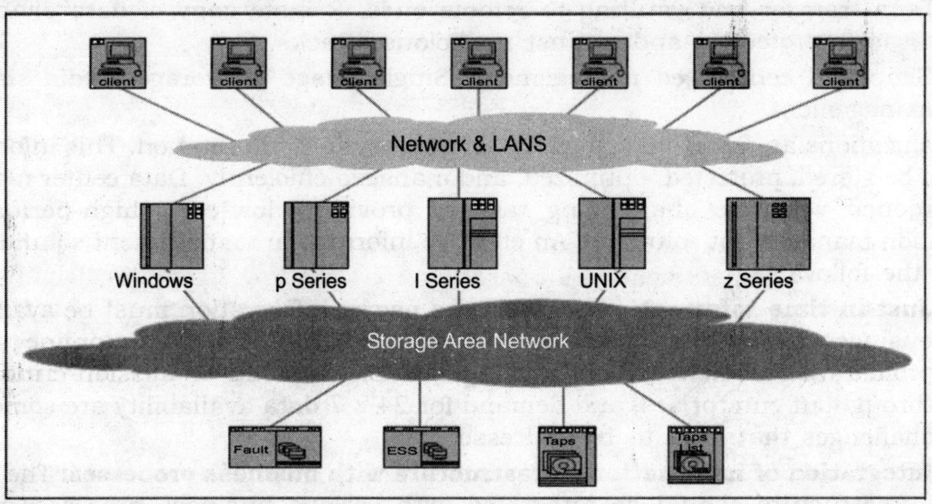

Fig. 4.5.

SANs create new methods of attaching storage to servers. These new methods can enable great improvements in both availability and performance. todays SANs are used to connect shared storage arrays and tape libaries to multiple servers, and are used by clustered servers for failover.

A SAN can be used to bypass traditional network bottlenecks. it facilitates direct, high-speed data transfers between servers and storage devices, potentially in any of the following three ways:

- **Server to storage:** This is the traditional model of interaction with storage devices. The advantage is that the same storage device may be accessed derially or concurrently by multiple servers.
- **Server to server:** A SAN may be used for high-speed, high-volume communications between servers.
- **Storage to storage:** This outboard data movement capability enables data to be moved without server intervention, thereby freeing up server processor cycles for other activities like application processing. Examples include a disk device backing up its data to a tape device without server intervention, or remote device mirroring across the SAN.

SANs allow applications that move data to perform better, for example, by having the data sent directly from the source to the target device with minimal server intervention. SANs also enable new network architectures where multiple hosts access multiple storage devices connected to the same network. Using a SAN can potentially offer the following benefits:

- Improvements to application availability; Storage is independent of applications and accessible through multiple data paths for better reliability, availability, and serviceability.
- Higher application performances: Storage processing is off-loaded from servers and moved onto a separate network.
- Centralized and consolidated storage: Simpler management, scalability, flexibility, and availability.

- Data transfer and vaulting to remote sites: Remote copy of data enabled for disaster protection and against mallicious attacks.
- Simplified centralized management: Single image of storage media simplifies management.

Organizations are experiencing an explosive growth in information. This information needs to be stored, protected, optimized, and managed efficiently. Data center managers are burdened with the challenging task of providing low-cost, high-performance information management solutions. An effective information management solution must provide the following:

- **Just-in-time information to business users:** Information must be available to business users when they need it. The explosive growth in online storage, proliferation of new servers and applications, spread of mission-critical data throughout enterprises, and demand for 24 × 7 data availability are some of the challenges that need to be addressed.
- **Integration of information infrastructure with business processes:** The storage infrastructure should be integrated with various business processes without compromising its security and integrity.
- **Flexible and resilient storage architecture:** The storage infrastructure must provide flexibility and resilience that aligns with changing business requirements. Storage should scale without compromising performance requirements of the applications and, at the same time, the total cost of managing information must be low.

Fibre Channel: Overview

The FC architecture forms the fundamental construct of the SAN infrastructure. *Fibre Channel* is a high-speed network technology that runs on high-speed optical fiber cables (preferred for front-end SAN connectivity) and serial copper cables (preferred for back-end disk connectivity). The FC technology was created to meet the demand for increased speeds of data transfer among computers, servers, and mass storage subsystems. Although FC networking was introduced in 1988, the FC standardization process began when the American National Standards Institute (ANSI) chartered the Fibre Channel Working Group (FCWG). By 1994, the new high-speed computer interconnection standard was developed and the Fibre Channel Association (FCA) was founded with 70 charter member companies. Technical Committee T11, which is the committee within INCITS (International Committee for Information Technology Standards), is responsible for Fibre Channel interfaces. T11 (previously known as X3T9.3) has been producing interface standards for high performance and mass storage applications since the 1970s.

Higher data transmission speeds are an important feature of the FC networking technology. The initial implementation offered throughput of 100 MB/s (equivalent to raw bit rate of 1Gb/s i.e. 1062.5 Mb/s in Fibre Channel), which was greater than the speeds of Ultra SCSI (20 MB/s) commonly used in DAS environments. FC in full-duplex mode could sustain throughput of 200 MB/s. In comparison with Ultra-SCSI, FC is a significant leap in storage networking technology.

Latest FC implementations of 8 GFC (Fibre Channel) offers throughput of 1600 MB/s (raw bit rates of 8.5 Gb/s), whereas Ultra320 SCSI is available with a throughput of 320 MB/s. The FC architecture is highly scalable and theoretically a single FC network can accommodate approximately 15 million nodes.

4.4.1. The SAN Evolution and Comparison

A *storage area network (SAN)* carries data between servers (also known as *hosts*) and storage devices through fibre channel switches (see Fig. 4.6.). A SAN enables storage consolidation and allows storage to be shared across multiple servers. It enables organizations to connect geographically dispersed servers and storage.

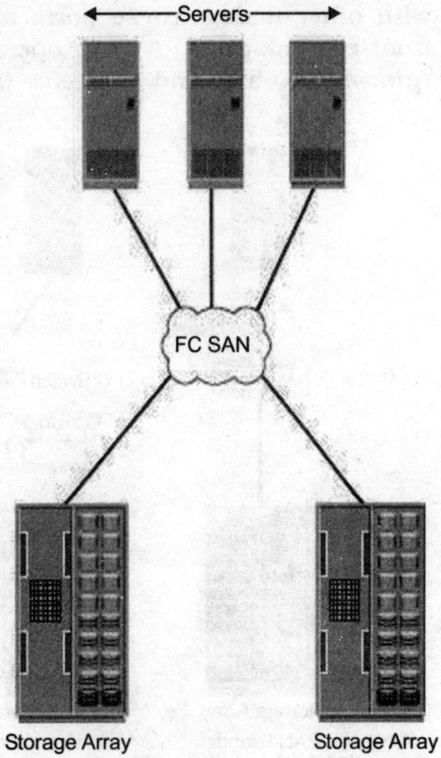

Fig. 4.6.

A SAN provides the physical communication infrastructure and enables secure and robust communication between host and storage devices. The SAN management interface organizes connections and manages storage elements and hosts.

In its earliest implementation, the SAN was a simple grouping of hosts and the associated storage that was connected to a network using a hub as a connectivity device. This configuration of a SAN is known as a *Fibre Channel Arbitrated Loop (FC-AL)*, which is detailed later in the chapter. Use of hubs resulted in isolated FC-AL SAN islands because hubs provide limited connectivity and bandwidth.

The inherent limitations associated with hubs gave way to high-performance FC *switches*. The switched fabric topologies improved connectivity and performance, which enabled SANs to be highly scalable. This enhanced data accessibility to applications across the enterprise. FC-AL has been abandoned for SANs due to its limitations, but still survives as a disk-drive interface.

4.4.1.1. Comparison

Direct-attached storage (DAS) is often referred to as a stove piped storage environment. Hosts "own" the storage and it is difficult to manage and share resources on these isolated storage devices. Efforts to organize this dispersed data led to the emergence of the storage area network (SAN). SAN is a highspeed, dedicated network of servers and shared storage devices. Traditionally connected over Fibre Channel (FC) networks, a SAN forms a single-storage pool and facilitates data centralization and consolidation. SAN meets the storage demands efficiently with better economies of scale. A SAN also provides effective maintenance and protection of data.

4.4.2. SAN and Its Application, Components

A SAN consists of three basic components: servers, network infrastructure, and storage. These components can be further broken down into the following key elements: node ports, cabling, interconnecting devices (such as FC switches or hubs), storage arrays, and SAN management software.

Node Ports

In fibre channel, devices such as hosts, storage and tape libraries are all referred to as *nodes*. Each node is a source or destination of information for one or more nodes. Each node requires one or more ports to provide a physical interface for communicating with other nodes. These ports are integral components of an HBA and the storage front-end adapters. A port operates in full-duplex data transmission mode with a *transmit (Tx)* link and a *receive (Rx)* link.

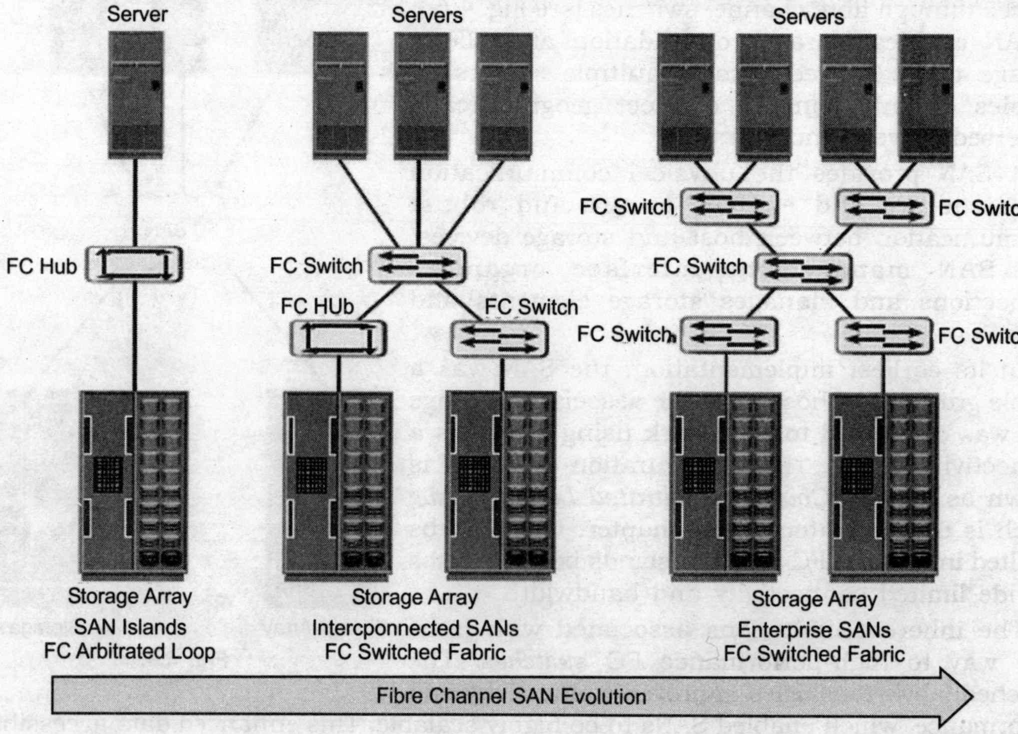

Fig. 4.7.

4.5. CONTENT-ADDRESSED STORAGE, ITS EVOLUTION AND COMPARISON

In the life cycle of information, data is actively created, accessed, edited, and changed. As data ages, it becomes less likely to change and eventually becomes "fixed" but continues to be accessed by multiple applications and users.

This data is called *fixed content*. Traditionally, fixed content was not treated as a specialized form of data and was stored using a variety of storage media, ranging from optical disks to tapes to magnetic disks. While these traditional technologies store content, none of them provide all of the unique requirements for storing and accessing fixed content.

Accumulations of fixed content such as documents, e-mail messages, web pages, and digital media throughout an organization have resulted in an unprecedented growth in the amount of data. It has also introduced the challenge of managing fixed content. Furthermore, users demand assurance that stored content has not changed and require an immediate online access to fixed content. These requirements resulted in the development of Content-Addressed Storage (CAS).

CAS is an *object-based system* that has been purposely built for storing fixed content data. It is designed for secure online storage and retrieval of fixed content.

Unlike file-level and block-level data access that use file names and the physical location of data for storage and retrieval, CAS stores user data and its attributes as separate objects. The stored object is assigned a globally unique address known as a *content address (CA)*. This address is derived from the object's binary representation. CAS provides an optimized and centrally managed storage solution that can support *single-instance storage (SiS)* to eliminate multiple copies of the same data.

4.5.1. CAS and Its Application, Components

Data is accessed and modified at varying frequencies between the time it is created and discarded. Some data frequently changes, for example, data accessed by an Online Transaction Processing (OLTP) application. Some data that does not typically change, but is allowed to change if required is, for example, bill of material and design documents.

Another category of data is fixed content, which defines data that cannot be changed. X-rays and pictures are examples of fixed content data. It is mandatary for all organizations to retain some data for an extended period of time due to government regulations and legal/contractual obligations. Fixed data, which is retained for future reference or business value, is referred to as fixed content asset. Some examples of fixed content asset include electronic documents, e-mail messages, Web pages, and digital media (see Fig. 4.8).

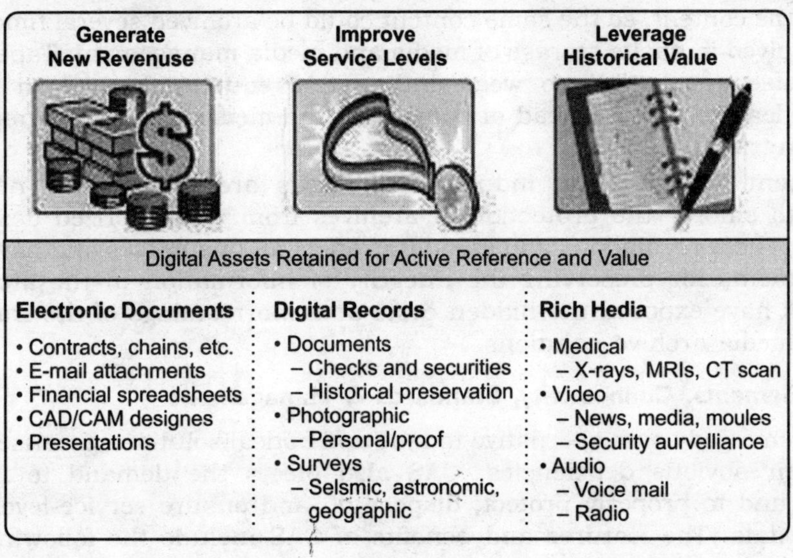

Fig. 4.8.

Organizations make use of these digital assets to generate new revenues, improve service levels, and leverage historical value. This demands frequent and quick retrieval of the fixed contents at multiple locations. An *archive* is a repository where fixed content is placed. Online data availability in the archive can further increase the business value of the referenced information.

4.5.2. CAS Security and Limitations

An electronic data archive is a repository for data that has fewer access requirements. It can be implemented as online, nearline, or offline based on the means of access:

- **Online archive:** The storage device is directly connected to the host to make the data immediately available. This is best suited for active archives.

- **Nearline archive:** The storage device is connected to the host and information is local, but the device must be mounted or loaded to access the information.
- **Offline archive:** The storage device is not directly connected, mounted, or loaded. Manual intervention is required to provide this service before information can be accessed.

An archive is often stored on a *write once read many (WORM)* device, such as a CD-ROM. These devices protect the original file from being overwritten. Some tape devices also provide this functionality by implementing filelocking capabilities in the hardware or software. Although these devices are inexpensive, they involve operational, management, and maintenance overhead.

Requirements to retain archives have caused corporate archives to grow at a rate of 50 percent or more per year. At the same time, organizations must reduce costs while maintaining required service-level agreements (SLAs).

Therefore, it is essential to find a solution that minimizes the fixed costs of the archive's operations and management.

Archives implemented using tape devices and optical disks involve many hidden costs. The traditional archival process using optical disks and tapes is not optimized to recognize the content, so the same content could be archived several times. Additional costs are involved in offsite storage of media and media management. Tapes and optical media are also susceptible to wear and tear. Frequent changes in these device technologies lead to the overhead of converting the media into new formats to enable access and retrieval.

Government agencies and industry regulators are establishing new laws and regulations to enforce the protection of archives from unauthorized destruction and modification. These regulations and standards affect all businesses and have established new requirements for preserving the integrity of information in the archives. These requirements have exposed the hidden costs and shortcomings of the traditional tape and optical media archive solutions.

4.5.3. CAS Elements, Connectivity, Standards & Management

CAS has emerged as an alternative to tape and optical solutions because it overcomes many of their obvious deficiencies. CAS also meets the demand to improve data accessibility and to properly protect, dispose of, and ensure service-level agreements for archived data. The features and benefits of CAS include the following:

- **Content authenticity:** It assures the genuineness of stored content. This is achieved by generating a unique content address and automating the process of continuously checking and recalculating the content address for stored objects. Content authenticity is assured because the address assigned to each piece of fixed content is as unique as a fingerprint. Every time an object is read, CAS uses a hashing algorithm to recalculate the object's content address as a validation step and compares the result to its original content address. If the object fails validation, it is rebuilt from its mirrored copy.
- **Content integrity:** Refers to the assurance that the stored content has not been altered. Use of hashing algorithm for content authenticity also ensures content integrity in CAS. If the fixed content is altered, CAS assigns a new address to the altered content, rather than overwrite the original fixed content, providing an audit trail and maintaining the fixed content in its original state. As an integral part of maintaining data integrity and audit trail capabilities,

CAS supports parity RAID protection in addition to mirroring. Every object in a CAS system is systematically checked in the background. Over time, every object is tested, guaranteeing content integrity even in the case of hardware failure, random error, or attempts to alter the content with malicious intent.

- **Location independence:** CAS uses a unique identifier that applications can leverage to retrieve data rather than a centralized directory, path names, or URLs. Using a content address to access fixed content makes the physical location of the data irrelevant to the application requesting the data. Therefore the location from which the data is accessed is transparent to the application. This yields complete content mobility to applications across locations.
- **Single-instance storage (SiS):** The unique signature is used to guarantee the storage of only a single instance of an object. This signature is derived from the binary representation of the object. At write time, the CAS system is polled to see if it already has an object with the same signature. If the object is already on the system, it is not stored, rather only a pointer to that object is created. SiS simplifies storage resource management tasks, especially when handling hundreds of terabytes of fixed content.
- **Retention enforcement:** Protecting and retaining data objects is a core requirement of an archive system. CAS creates two immutable components: a data object and a meta-object for every object stored. The metaobject stores object's attributes and data handling policies. For systems that support object-retention capabilities, the retention policies are enforced until the policies expire.
- **Record-level protection and disposition:** All fixed content is stored in CAS once and is backed up with a protection scheme. The array is composed of one or more storage clusters. Some CAS architectures provide an extra level of protection by replicating the content onto arrays located at a different location. The disposition of records also follows the stringent guidelines established by regulators for shredding and disposing of data in electronic formats.
- **Technology independence:** The CAS system interface is impervious to technology changes. As long as the application server is able to map the original content address the data remains accessible. Although hardware changes are inevitable, the goal of CAS hardware vendors is to ensure compatibility across platforms.
- **Fast record retrieval:** CAS maintains all content on disks that provide subsecond "time to first byte" (200 ms–400 ms) in a single cluster. Random disk access in CAS enables fast record retrieval.

4.5.4. CAS Architecture

The CAS architecture is shown in Fig. 4.9. A client accesses the CAS-Based storage over a LAN through the server that runs the CAS API (application programming interface). The CAS API is responsible for performing functions that enable an application to store and retrieve the data. CAS architecture is a *Redundant Array of Independent Nodes (RAIN).* It contains storage nodes and access nodes networked as a cluster by using a private LAN that is internal to it. The internal LAN can be reconfigured automatically to detect the configuration changes such as the addition of storage or access nodes.

Clients access the CAS on a separate LAN, which is used for interconnecting clients and servers to the CAS.

The nodes are configured with low-cost, high-capacity ATA HDDs. These nodes run an operating system with special software that implements the features and functionality required in a CAS system.

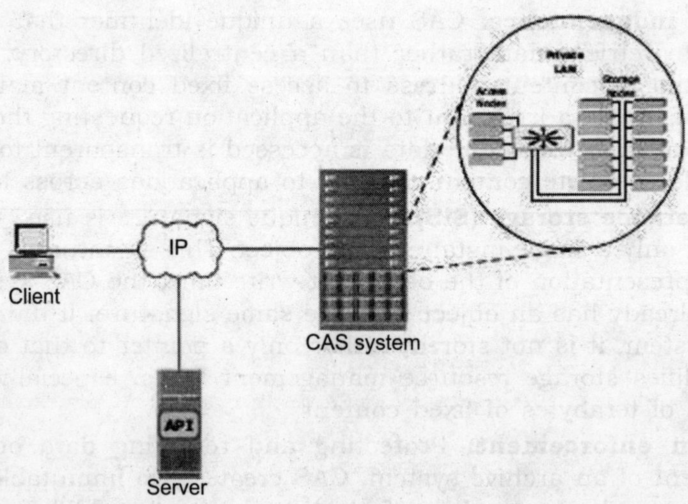

Fig. 4.9.

When the cluster is installed, the nodes are configured with a "role" defining the functionality they provide to the cluster. A node can be configured as a storage node, an access node, or a dual-role node. *Storage nodes* store and protect data objects. They are sometimes referred to as *back-end nodes. Access nodes* provide connectivity to application servers through the customer's LAN. They establish connectivity through a private LAN to the storage nodes in the cluster. The number of access nodes is determined by the amount of user required throughput from the cluster. If a node is configured solely as an "access node," its disk space cannot be used to store data objects. This configuration is generally found in older installations of CAS. Storage and retrieval requests are sent to the access node via the customer's LAN. *Dual-role nodes* provide both storage and access node capabilities. This node configuration is more typical than a pure access node configuration.

Almost all CAS products have the same features and options. Some may be implemented differently, but the following features are an essential part of any CAS solution:

- **Integrity checking:** It ensures that the content of the file matches the digital signature (hashed output or CA). The integrity checks can be done on every read or by using a background process. If problems are identified in any of the objects the nodes automatically repair or regenerate the object.
- **Data protection and node resilience:** This ensures that the content stored on the CAS system is available in the event of disk or node failure. Some CAS systems provide local replication or mirrors that copy a data object to another node in the same cluster. This decreases the total available capacity by 50 percent. Parity protection is another way to protect CAS data. It uses less capacity to store data, but takes longer to regenerate the data if corrupted. Remote replication copies data objects to a secondary storage device in a remote location. Remote replication is used as a disasterrecovery solution or for backup. Replication technologies are detailed in Chapters 13 and 14.

- **Load balancing:** Distributes data objects on multiple nodes to provide maximum throughput, availability, and capacity utilization.
- **Scalability:** Adding more nodes to the cluster without any interruption to data access and with minimum administrative overhead.
- **Self diagnosis and repair:** Automatically detects and repairs corrupted objects and alert the administrator of any potential problem. These failures can be at an object level or a node level. They are transparent to the users who access the archive. CAS systems can be configured to alert remote support teams who diagnose and make repairs remotely.
- **Report generation and event notification:** Provides on-demand reporting and event notification. A command-line interface (CLI) or a graphical user interface (GUI) can produce various types of reports. Any event notification can be communicated to the administrator through syslog, SNMP, SMTP, or e-mail.
- **Fault tolerance:** Ensures data availability if a component of the CAS system fails, through the use of redundant components and data protection schemes. If remote replication of CAS is implemented, failover to the remote CAS system occurs when the primary CAS system is unavailable.
- **Audit trails:** Enable documentation of management activity and any access and disposition of data. Audit trails are mandated by compliance requirements.

The data object storage process in a CAS system is as follows:

1. End users present the data to be archived to the CAS API via an application. The application server may also interact directly with the source (e.g., an X-ray machine) that generated this fixed content.
2. The API separates the actual data (BLOB) from the metadata and the CA is calculated from the object's binary representation.
3. The content address and metadata of the object are then inserted into the C-Clip Descriptor File (CDF). The C-clip is then transferred to and stored on the CAS system.
4. The CAS system recalculates the object's CA as a validation step and stores the object. This is to ensure that the content of the object has not changed.
5. An acknowledgment is sent to the API after a mirrored copy of the CDF and a protected copy of the BLOB have been safely stored in the CAS system. After a data object is stored in the CAS system, the API is given a C-Clip ID and C-Clip ID is stored local to the application server.
6. Using the C-Clip ID, the application can read the data back from the CAS system.

Once an object is stored successfully, it is made available to end users for retrieval and use. The content address is usually hidden from the user. A user accesses the file stored on CAS by the same file name. It is the application server that references the CA to retrieve the stored content; this process is transparent to the user. No modification is needed to the application's user interface to accommodate the CAS storage and retrieval process.

Application Server

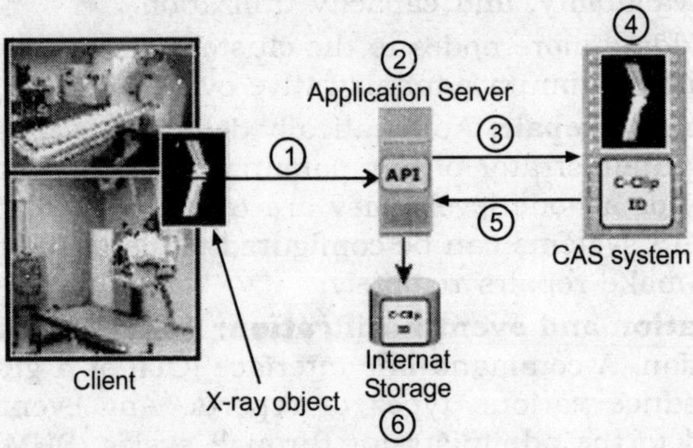

Fig. 4.10.

4.6. STANDARDS, MANAGEMENT, SECURITY AND LIMITATIONS OF DAS, NAS, CAS AND SAN

Direct attached storage (DAS) The simplest type of data storage is called direct attached storage (DAS). In a DAS configuration one or more data storage components, such as hard disks or tape drives, are either installed in a computer or connected directly to it – often with a SCSI or SAS link. One emerging storage option is shared DAS, in which a fixed number of servers are connected directly to a storage system instead of using a storage network, or fabric.

In direct attached storage (DAS), each server is configured with its own separate storage.

Example of a real-world DAS solution The HP StorageWorks 60 Modular Smart Array (MSA60) is an example of DAS storage. This 12-drive disk enclosure supports both SATA and SAS hard drives and offers quick and easy expansion for individual servers.

Example of shared DAS

The HP StorageWorks 2000sa Modular Smart Array (MSA2000sa) allows up to four servers to share a common pool of disks. The MSA2000sa supports data protection features such as snapshots and cloning, and it can be expanded with three additional drive shelves.

DAS in brief

What it is	Storage in or attached diectly to a server
Advantages	• Low initial cost and easy to implement • Does not require storage expertise; managed by the servers operating system
Disadvantages	• Servers cannot share storage capacity, which causes 'islands' of capacity - one servers storage is at 99 percent use, another at 5 per cent use • Does not scale up well • Difficult to manage if more than a couple of servers

Use DAS if ...	• Immediate start-up cost is your only consideration • You have no plans to grow in the future
Does not use it if ...	• You have many servers to be interconnected with centralized storage

Shared DAS in brief

What it is	Shared storage connected directly to a fixed, small number of servers
Advantages	• Reduced cost and complexity because there is no storage networking infrastructure • Easy to deploy and manage
Disadvantages	• Does not scale beyond a small number of servers • Limited distance between servers and storage
Use if ...	• You are setting up a small application cluster with two to four servers
Do not use it if ...	• More than four servers need shared storage • You need to connect other devices such as tape to a switched storage network

Network Attached Storage (NAS)

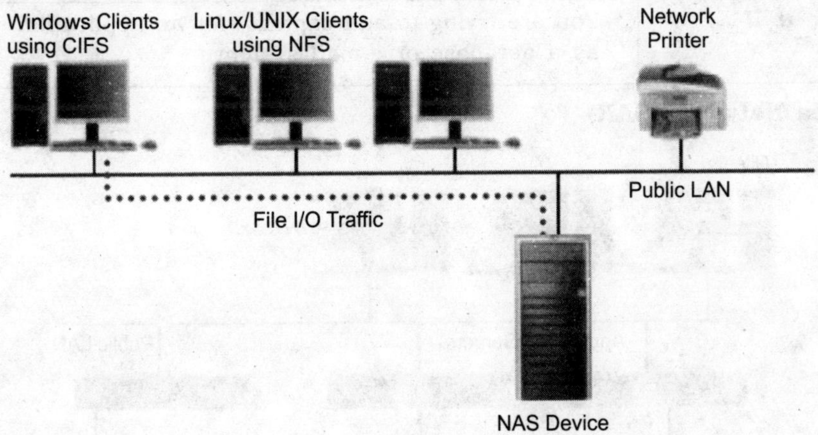

Fig. 4.11.

Network attached storage (NAS)

Network attached storage (NAS) is essentially a dedicated file server running an operating system that is designed and tuned specifically to handle file I/O traffic for network clients.

Client computer access to the NAS server is usually over an Ethernet connection. The NAS server appears on the network as a single node, with its own IP address. Files stored on the NAS system are accessible to clients on the LAN over the Ethernet connection via protocols such as CIFS/SMB (Windows® clients) or NFS (Linux and UNIX®). Many NAS systems also support protocols such as HTTP or FTP for Internet-based file access.

NAS products range from low-cost home office– quality devices up to fully clustered enterprise NAS gateways that connect to a SAN. In network attached storage (NAS), all clients have access to the same storage via the local area network (LAN).

Example of a real-world NAS solution The HP ProLiant ML110 G5 Storage Server is a NAS file server solution for small offices with less than 50 employees. This solution draws on your skill set and investment in ProLiant servers and Windows Server operating systems.

NAS in brief

What it is	Dedicated storage server attached to local area network, running OS tuned specifically for file serving
I/O	File I/O
Advantages	• Simple and economical; low TCO • Options for small business through enterprise • Works well in multi-OS environments (Windows, Linux, Macintosh...) • On-demand assignment of file shares to clients and servers
Disadvantages	• Uses public LAN; can still have bottlenecks • Server device can be single point of failure
Use NAS if ...	• You need economical, easy-to-administer file serving over the existing network infrastructure
Do not use it if ...	• You are trying to add capacity for an application server such as a database or e-mail system

Storage Area Network (SAN)

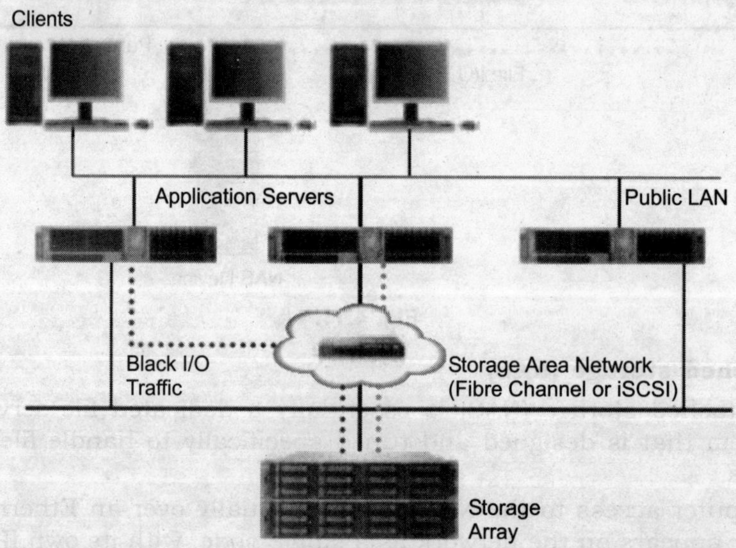

Fig. 4.12.

A storage area network (SAN) is just that—a network that is dedicated to storage. The SAN is separate from the LAN and provides servers with access to storage. The network is designed to handle storage communications.

In its simplest form, a SAN consists of: Shared storage (• typically a disk array)

- The dedicated network that lets servers and storage 'talk' to each other
- Software that configures the shared environment
- Components to protect data in the SAN

To ensure optimal performance, the network between storage and servers is typically Fibre Channel, multi-gigabit iSCSI, or another high-speed link.

A storage area network (SAN) is a high-speed network—only for storage—that is separate from the public LAN.

Example of a real-world SAN solution. The HP StorageWorks 2000i Modular Smart Array (MSA2000i) enables up to 16 servers to share a common pool of disks across an Ethernet network. The MSA2000i supports data protection features such as snapshots and cloning and can be expanded with three additional drive shelves.

SAN Storage in brief

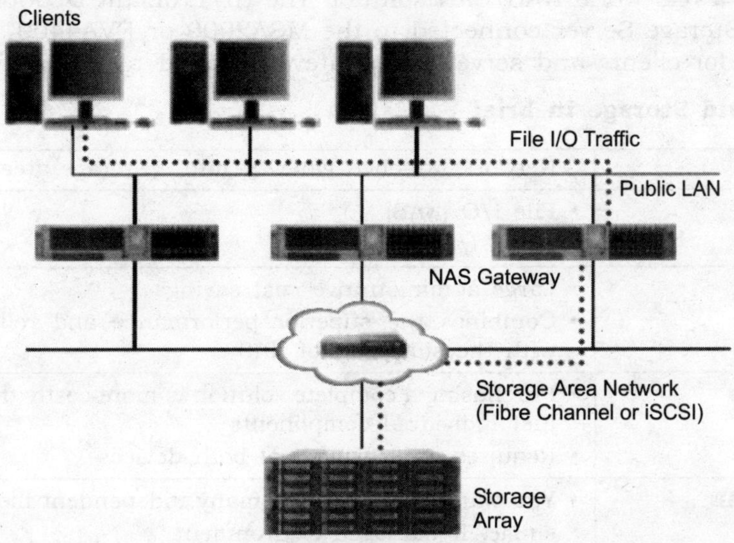

Fig. 4.13.

What it is	Separate network dedicated to storage connecting servers to external disk subsystems
I/O	Block I/O
Advantages	• Multiple servers can share common pool of storage • More efficient and flexible capacity utilization for lower TCO • Improved scalability, reliability and resource management
Disadvantages	• More complex to set up than a basic LAN • Often requires knowledge of storage configuration and management • Does not support file sharing directly; instead uses a file server or NAS connected to the SAN
Use a SAN if ...	• You need the fastest data access speeds through a network such as Fibre Channel • You want a backup solution that reduces server and LAN utilization

Do not use if if ...	• You have only a single server and a few workstations • Cost is the primary decision factor AS gateway with a storage area network (SAN)

NAS gateway with a SAN

One common use of a storage area network is to store file data, providing access to that data via a NAS gateway. The gateway has no on-board storage of its own; instead, it connects to a SAN and acts as a translator between the file-level NAS protocols, such as NFS and CIFS, and the block-level SAN protocols such as Fibre Channel and iSCSI.

This NAS/SAN hybrid combines the advantages of both technologies. Using a NAS gateway to a SAN combines the advantages of both technologies, providing both file and block I/O.

Example of a real-world NAS/SAN solution The HP ProLiant DL380 G5 or SB460c SAN Gateway Storage Server connected to the MSA2000 or EVA4400, providing easy access to data for clients and servers in whatever method required—file or block.

NAS/SAN hybrid Storage in brief

What it is	Network attached gateway into a storage area network
I/O	• File I/O (NAS) • Block I/O (SAN)
Advantages	• Large administrative cost savings • Combines the superior performance and reliability of SAN with the simplicity of NAS
Disadvantages	• Purchasing a complete solution is more costly than purchasing just individual components • Requires management of both devices
Use a NAS/SAN hybrid if ...	• You want to consolidate many independent file servers into a single, It managed environment • You want equal access to file and block-based data
Do not use it if ...	• A small, simple storage solution is required.

SUMMARY

1. Computing is based on information. Information is the underlying resource on which all computing processes are based; it is a company asset. Information is stored on storage media, and is accessed by applications executing on a server. Often the information is a unique company asset. Information is created and acquired every second of every day.
2. Information is the currency of business. To ensure that any business delivers the expected results, they must have access to accurate information, and without delay.
3. The management and protection of business information is vital for the availability of business processes. This chapter introduces the concept of a Storage Area Network, which has been regarded as the ultimate response to all these needs.
4. A SAN can also be a storage system consisting of storage elements, storage devices, computer systems, and/or appliances, plus all control software, communicating over a network. Put in simple terms, a SAN is a specialized, high-speed network attaching

servers and storage devices and, for this reason, It is sometimes referred to as "the network behind the servers."

5. A SAN allows "any-to-any" connection across the network, using interconnect elements such as routers, gateways, hubs, switches and directors. It eliminates the traditional dedicated connection between a server and storage, and the concept that the server effectively "owns and manages" the storage devices.
6. It also eliminates any restriction to the amount of data that a server can access, currently limited by the number of storage devices attached to the individual server.
7. Instead, a SAN introduces the flexibility of networking to enable one server or many heterogeneous servers to share a common storage utility, which may comprise many storage devices, including disk, tape, and optical storage. Additionally, the storage utility may be located far from the servers that use it.
8. The SAN can be viewed as an extension to the storage bus concept, which enables storage devices and servers to be interconnected using similar elements as in local area networks (LANs) and wide area networks (WANs): Routers, hubs, switches, directors, and gateways.
9. A SAN can be shared between servers and/or dedicated to one server. It can be local, or can be extended over geographical distances.
10. Disk storage is often purchased from the processor vendor as an integral feature, and it is difficult to establish if the price you pay per gigabyte (GB) is competitive, compared to the market price of disk storage. Disks and tape drives, directly attached to one client or server, cannot be used by other systems, leading to inefficient use of hardware resources.

QUESTION WITH ANSWER

Q.1. What is Networked Storage (SAN)?

Ans. The Storage Network Industry Association (SNIA) defines the SAN as a network whose primary purpose is the transfer of data between computer systems and storage elements. A SAN consists of a communication infrastructure, which provides physical connections; and a management layer, which organizes the connections, storage elements, and computer systems so that data transfer is secure and robust. The term SAN is usually (but not necessarily) identified with block I/O services rather than file access services.

A SAN can also be a storage system consisting of storage elements, storage devices, computer systems, and/or appliances, plus all control software, communicating over a network. Put in simple terms, a SAN is a specialized, high-speed network attaching servers and storage devices and, for this reason, It is sometimes referred to as "the network behind the servers." A SAN allows "any-to-any" connection across the network, using interconnect elements such as routers, gateways, hubs, switches and directors. It eliminates the traditional dedicated connection between a server and storage, and the concept that the server effectively "owns and manages" the storage devices. It also eliminates any restriction to the amount of data that a server can access, currently limited by the number of storage devices attached to the individual server. Instead, a SAN introduces the flexibility of networking to enable one server or many heterogeneous servers to share a common storage utility, which may comprise many storage devices, including disk, tape, and optical storage. Additionally, the storage utility may be located far from the servers that use it.

Note: The SNIA definition specifically does not identify the term SAN with Fibre Channel technology. When the term SAN is used in connection with Fibre Channel technology, use of a qualified phrase such as Fibre Channel SAN is encouraged. According to this definition, an Ethernet-based networkwhose primary purpose is to provide access to storage elements would be considered a SAN. SANs are sometimes also used for system interconnection in clusters.

The diagram in Fig. 4.14, shows a tiered overview of a SAN connecting multiple servers to multiple storage systems.

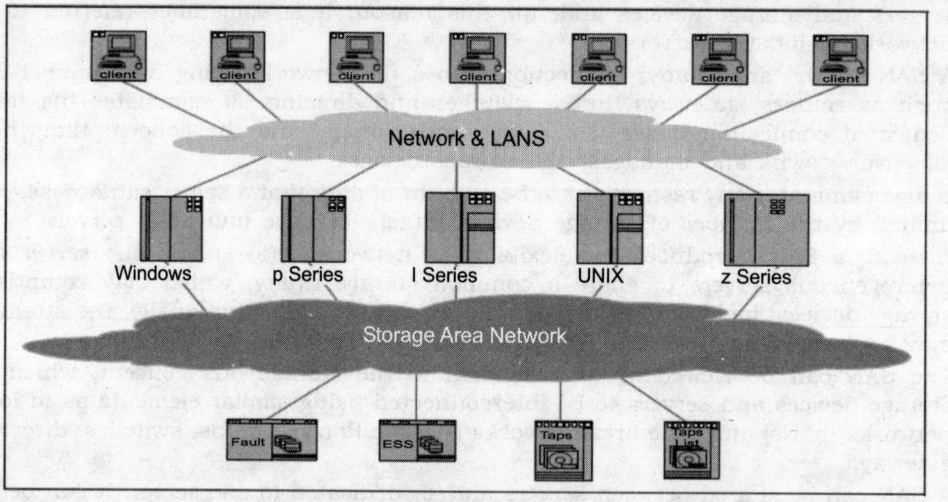

Fig. 4.14.

SANs create new methods of attaching storage to servers. These new methods can enable great improvements in both availability and performance. todays SANs are used to connect shared storage arrays and tape libaries to multiple servers, and are used by clustered servers for failover.

Q.2. Why use a SAN?

Ans. In this section we describe the main motivators that drive SAN implementations, and present some of the key benefits that this technology might bring to data-dependent business.

The problem

As illustrated in Figure C-2, the 1990's witnessed a major shift away from the traditional mainframe, host- centric model of computing to the client/server model. Today, many organizations have hundreds, even thousands, of distributed servers and client systems installed throughout its IT infrastructure. Many of these systems are powerful computers, with more processing capability than many mainframe computers had only a few years ago.

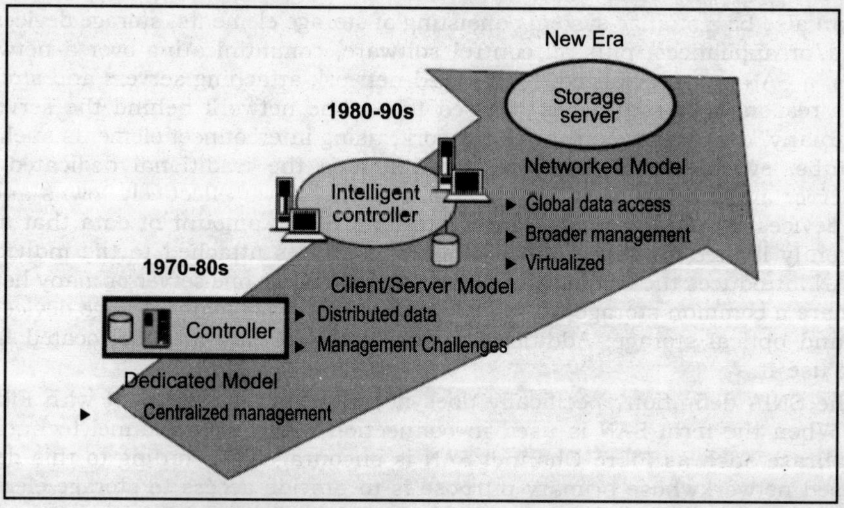

Fig. 4.15. The evolution of storage architecture

Storage, for the most part, is directly connected by a dedicated channel to the server it supports. Frequently the servers are interconnected using local area networks (LAN) and wide area networks (WAN), to communicate and exchange data. The amount of disk storage capacity attached to such systems has grown exponentially in recent years. It is a common place for a desktop personal computer or ThinkPad today to have storage in the order of tens of gigabytes. There has been a move to disk arrays, comprising a number of disk drives. The arrays may be "just a bunch of disks" (JBOD), or various implementations of redundant arrays of independent disks (RAID). The capacity of such arrays may be measured in tens or hundreds of gigabytes, but I/O bandwidth has not kept pace with the rapid growth in processor speeds and disk capacities.

Q.3 How can we use a SAN?

Ans. The key benefits that a SAN might bring to a highly data-dependent business infrastructure can be summarized into three rather simple concepts: infrastructure simplification, information lifecycle management and business continuity. They are an effective response to the requirements presented in the previous section, and are strong arguments for the adoption of SANs.

These three concepts are briefly described as follows:

A Infrastructure simplification: There are four main methods by which infrastructure simplification can be achieved: *consolidation, virtualization, automation and integration*:

1. **Consolidation:** Concentrating systems and resources into locations with fewer, but more powerful, servers and storage pools can help increase IT efficiency and simplify the infrastructure. Additionally, centralized storage management tools can help improve scalability, availability, and disaster tolerance.
2. **Virtualization;** Storage virtualization helps in making complexity nearly transparent and at the same time can offer a composite view of storage assets. This may help reduce capital and administrative costs, while giving users better service and availability. Virtualization is designed to help make the IT infrastructure more responsive, scalable, and available.
3. **Automation:** Choosing storage components with autonomic capabilities can improve availability and responsiveness-and help protect data as storage needs grow. As soon as day-to-day tasks are automated, storage administrators may be able to spend more time on critical, higher-level tasks unique to a company's business mission.
4. **Integration:** Integrated storage environments simplify system management tasks and improve security. When all servers have secure access to all data, your infrastructure may be better able to respond to your users information needs.

Q.4. Define JBOD, DAS, NAS, SAN & CAS evolution and comparision

Ans. This section briefly describes the main types of storage devices that can be found in the market.

Direct attached storage (DAS): The simplest type of data storage, located in or attached directly to a server.

- *Storage area network (SAN):* A dedicated network for storage traffic between servers and a disk storage array or tape device.
- *Hybrid or unified NAS/SAN (NAS gateway):* A combination of a NAS gateway with a SAN and a hybrid NAS/SAN, with a simple management interface to hide complexity.

Disk systems: In brief a disk system is a device in which a number of physical storage disks sit side-by-side. By being contained within a single "box", a disk system usually has a central control unit that manages all the I/O, simplifying the integration of the system with other devices, such as other disk systems or servers.

Depending on the "intelligence" with which this central control unit is able to manage the individual disks, a disk system can be a JBOD or a RAID.

Just A Bunch Of Disks (JBOD): In this case, the disk system appears as a set of individual storage devices to the device they are attached to. The central control unit provides only basic functionality for writing and reading data from the disks.

Redundant Array of Independent Disks (RAID): In this case, the central control unit provides additional functionality that makes it possible to utilize the individual disks in such a way to achieve higher fault-tolerance and/or performance. The disks themselves appear as a single storage unit to the devices to which they are connected.

Depending on the specific functionality offered by a particular disk system, it is possible to make it behave as a RAID and/or a JBOD; the decision as to which type of disk system is more suitable for a SAN implementation strongly depends on the performance and availability requirements for this particular SAN.

Q.5. What are the Applications, Elements & connectivity of DAS, NAS, CAS & SAN

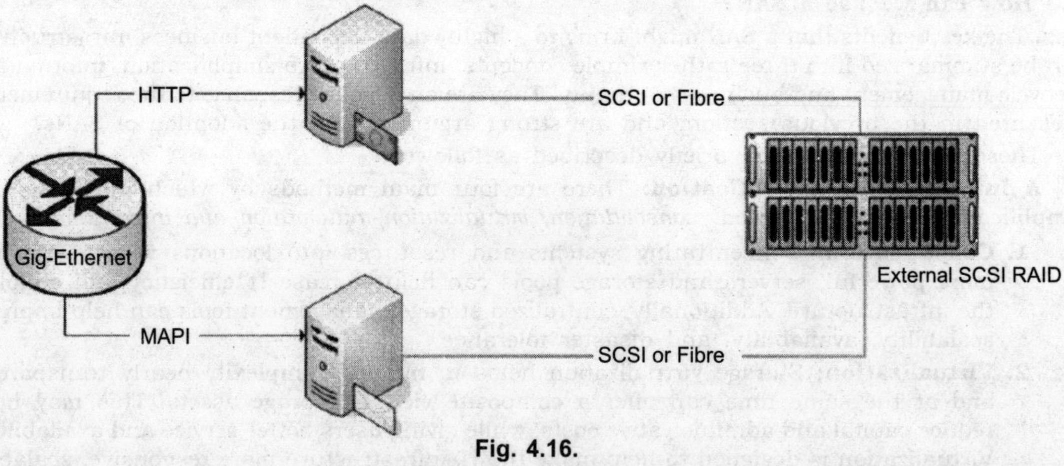

Fig. 4.16.

Ans. It didn't take long to see the appearance of disk cabinets-or Disk Arrays, connected to servers via an external SCSI cable, that were separately managed. In some cases these storage cabinets could be connected to multiple servers, so they could share the storage (perhaps for fault tolerance). Also, being able to 'hot swap' failed disks and have the unit rebuild that disk from parity on the other disks was an expected feature.

This lead to the acronym DAS, or Directly Attached Storage (actually the acronym was coined in more recent times to distinguish it from other technologies). The main technologies used with DAS are SCSI with a specialized Host Bus Adapters (HBA) installed in the servers. (More on HBAs later).

A DAS afforded multiple server-access (up to 4), for clustering but the main disadvantage was that DAS ended up yielding an island of information.

Q.6. Explain Network Storage Fault tolerance.

Ans. Multipath I/O is a fault-tolerance and performance enhancement technique whereby there is more than one physical path between a computer system and its mass storage devices through the buses, controllers, switches, and bridge devices connecting them.

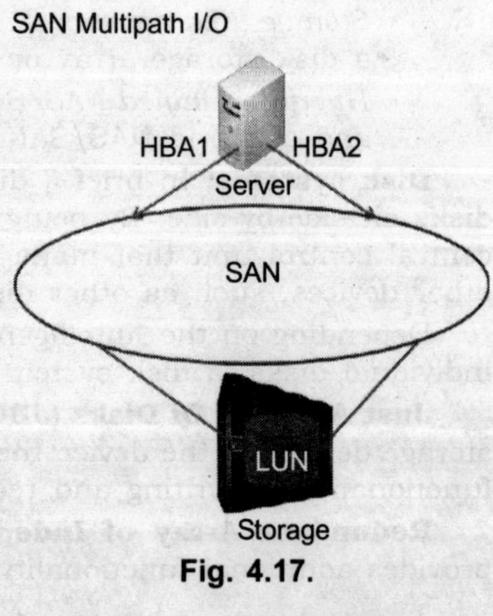

Fig. 4.17.

In a well-designed SAN, it is likely that you will want a device to be accessed by the host application over more than one path in order to potentially obtain better performance, and to facilitate recovery in the case of adapter, cable, switch, or GBIC failure. Should one controller, port or switch fail, the servers OS can route I/O through the remaining controller transparently to the application, with no changes visible to the applications, other than perhaps incremental latency. But, the same logical volume within a storage device (LUN) may be presented many times to the server through each of the possible paths to that LUN. In order to avoid this and make the device easier to administrate and to eliminate confusion, multipathing software is needed. This is responsible for making each LUN visible only once from the application and OS point of view. In addition to this, the multipathing software is also responsible for failover recovery and load balancing:

- **Failover recovery:** In a case of the malfunction of a component involved in making the LUN connection, the multipathing software redirects all the data traffic onto other available paths.
- **Load balancing:** The multipathing software is able to balance the data traffic equitably over the available paths from the hosts to the LUNs.

Q.7. What do you mean by Storage Replication.

Fig. 4.18.

Ans. Depending on the details behind how the particular replication works, the application layer may or may not be involved. If blocks are replicated without the knowledge of file systems or applications built on top of the blocks being replicated, when recovering using these blocks, the file system may be in an inconsistent state.

- A "Restartable" recovery implies that the application layer has full knowledge of the replication, and so the replicated blocks that represent the applications are in a consistent state. This means that the application layer (and possibly OS) had a chance to "quiesce" before the replication cycle.
- A "Recoverable" recovery implies that some extra work needs to be done to the replicated data before it can be useful in a recovery situation.

Q.8. What are the Standards, management, security and limitations of DAS?

Ans. Direct attached storage (DAS) The simplest type of data storage is called direct attached storage (DAS). In a DAS configuration one or more data storage components, such as hard disks or tape drives, are either installed in a computer or connected directly to it–often with a SCSI or SAS link. One emerging storage option is shared DAS, in which a fixed number of servers are connected directly to a storage system instead of using a storage network, or fabric. In direct attached storage (DAS), each server is configured with its own separate storage.

Example of a real-world DAS solution The HP StorageWorks 60 Modular Smart Array (MSA60) is an example of DAS storage. This 12-drive disk enclosure supports both SATA and SAS hard drives and offers quick and easy expansion for individual servers.

Example of shared DAS The HP StorageWorks 2000sa Modular Smart Array (MSA2000sa) allows up to four servers to share a common pool of disks. The MSA2000sa supports data protection features such as snapshots and cloning, and it can be expanded with three additional drive shelves.

DAS in brief

What it is	Storage in or attached diectly to a server
Advantages	• Low initial cost and easy to implement • Does not require storage expertise; managed by the servers operating system

Disadvantages	• Servers cannot share storage capacity, which causes 'islands' of capacity—one servers storage is at 99 percent use, another at 5 percent use • Does not scale up well • Difficult to manage if more than a couple of servers
Use DAS if ...	• Immediate start-up cost is your only consideration • You have no plans to grow in the future
Does not use it if ...	• You have many servers to be interconnected with centralized storage

Shared DAS in brief

What it is	Shared storage connected directly to a fixed, small number of servers
Advantages **Disadvantages**	• Reduced cost and complexity because there is no storage networking infrastructure • Easy to deploy and manage • Does not scale beyond a small number of servers • Limited distance between servers and storage
Use if ...	• You are setting up a small application cluster with two to four servers
Do not use it if ...	• More than four servers need shared storage • You need to connect other devices such as tape to a switched storage network

Q.9. What are the Standards, management, security and limitations of NAS ?

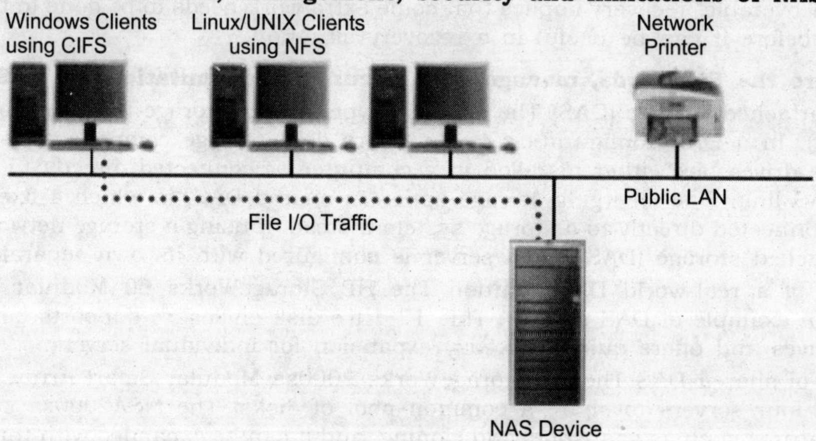

Fig. 4.19.

Ans. Network attached storage (NAS) Network attached storage (NAS) is essentially a dedicated file server running an operating system that is designed and tuned specifically to handle file I/O traffic for network clients. Client computer access to the NAS server is usually over an Ethernet connection. The NAS server appears on the network as a single node, with its own IP address. Files stored on the NAS system are accessible to clients on the LAN over the Ethernet connection via protocols such as CIFS/SMB (Windows clients) or NFS (Linux and UNIX®). Many NAS systems also support protocols such as HTTP or FTP for Internet-based file access.

NAS products range from low-cost home office- quality devices up to fully clustered enterprise NAS gateways that connect to a SAN. In network attached storage (NAS), all clients have access to the same storage via the local area network (LAN).

Example of a real-world NAS solution The HP ProLiant ML110 G5 Storage Server is a NAS file server solution for small offices with less than 50 employees. This solution draws on your skill set and investment in ProLiant servers and Windows Server operating.

OBJECTIVE TYPE QUESTIONS

1. A customer has a DMX-3 system with available low cost Fiber Channel storage and wants to use it as a backup-to-disk destination storage device. Some of the storage nodes in the network are not connected to the SAN.

   Which solution can be used so that all storage nodes in the environment use the DMX 3 as the backup-to-disk destination device while leveraging their current investment?

   (a) CX80G (b) B.DL210
   (c) NS80 (d) NS80G

2. A customer wants to expand an existing Symmetrix DMX-3 production system for use with backup-to-disk. What do you recommend?

   (a) Add ATA disks (b) Dedicate a quadrant to backup
   (c) Use Fibre Channel disks only (d) Use RAID 3 with the backup LUNs

3. In an EMC Disk Library (EDL) environment, what describes the auto-archive feature?

   (a) Backup application issues clone command; EDL copies save sets to physical tape with a different barcode
   (b) Backup application issues clone command; EDL copies save sets to physical tape with the same barcode
   (c) Backup application issues eject command; EDL moves backup data to physical tape with a different barcode
   (d) Backup application issues eject command; EDL moves backup data to physical tape with the same barcode

4. You designed a customer solution using a disk device for backups. The available disk space is limited. You determined that it is necessary to move data from disk to tape after a specified time. What should you use?

   (a) Advanced file type device with cloning
   (b) Advanced file type device with staging
   (c) File type device and nsrcompress
   (d) File type device with archiving

5. You designed a customer solution using a disk device for backups. The available disk space is limited. You determined that it is necessary to move data from disk to tape after a specified time. What should you use?

   (a) Advanced file type device with cloning
   (b) Advanced file type device with staging
   (c) File type device and nsrcompress
   (d) File type device with archiving

ANSWERS

1. (d) 2. (b) 3. (b) 4. (d) 5. (b)

CHAPTER

5

HYBRID STORAGE SOLUTIONS

5.0. INTRODUCTION

Corporations Are Experiencing An Unprecedented Combination Of Data Growth, Economic Ressure, And Regulatory Compliance. Managing Storage Costs And Meeting Service Levels Is Becoming Increasingly Difficult With Current Diskand Tape-based Solutions.

Fixed content, also known as persistent data, is the largest and fastest growing data type in the enterprise. Leading analysts stimate that it accounts for as much as 80% of unstructured data and is growing at approximately 100% CAGR. This data is not changing, not mission critical,and infrequently accessed; however, much of the data must be retained and retrieved quickly to satisfy increasing regulatory and legal requirements. In order to meet the online requirements, fixed content remains intermingled with production data on expensive primary storage arrays and is the culprit behind a myriad of inefficiencies that are rippling through data centers worldwide.

Inside this Chapter

5.1. VIRTUALIZATION

Storage networking technology matures, larger and complex implementations are becoming more common. The heterogeneous nature of storage infrastructures has further added to the complexity of managing and utilizing storage resources effectively.

Specialized technologies are required to meet stringent service level agreements and to provide an adaptable infrastructure with reduced cost of management. The virtualization technologies discussed in this chapter provide enhanced productivity, asset utilization, and better management of the storage infrastructure.

Virtualization is the technique of masking or abstracting physical resources, which simplifies the infrastructure and accommodates the increasing pace of business and technological changes. It increases the utilization and capability of IT resources, such as servers, networks, or storage devices, beyond their physical limits. Virtualization simplifies resource management by pooling and sharing resources for maximum utilization and makes them appear as logical resources with enhanced capabilities.

Virtualization has existed in the IT industry for several years and in differentforms, including memory virtualization, network virtualization, server virtualization, and storage virtualization

5.1.1. Memory Virtualization

Virtual memory makes an application appear as if it has its own contiguous logical memory independent of the existing physical memory resources.

Since the beginning of the computer industry, memory has been and continues to be an expensive component of a host. It determines both the size and the number of applications that can run on a host.

With technological advancements, memory technology has changed and the cost of memory has decreased. Virtual memory managers (VMMs) have evolved, enabling multiple applications to be hosted and processed simultaneously.

In a virtual memory implementation, a memory address space is divided into contiguous blocks of fixed-size pages. A process known as *paging* saves inactive memory pages onto the disk and brings them back to physical memory when required. This enables efficient use of available physical memory among different processes. The space used by VMMs on the disk is known as a *swap file.* A swap file (also known as *page file* or *swap space*) is a portion of the hard disk that functions like physical memory (RAM) to the operating system. The operating system typically moves the least used data into the swap file so that RAM will be available for processes that are more active. Because the space allocated to the swap file is on the hard disk (which is slower than the physical memory), access to this file is slower.

5.1.2. Network Virtualization

Network virtualization creates virtual networks whereby each application sees its own logical network independent of the physical network. A *virtual LAN (VLAN)* is an example of network virtualization that provides an easy, flexible, and less expensive way to manage networks. VLANs make large networks more manageable by enabling a centralized configuration of devices located in physically diverse locations.

Consider a company in which the users of a department are separated over a metropolitan area with their resources centrally located at one office. In a typical network, each location has its own network connected to the others through routers.

When network packets cross routers, latency influences network performance. With VLANs, users with similar access requirements can be grouped together into the same virtual network. This setup eliminates the need for network routing. As a result, although users are physically located at disparate locations, they appear to be at the same location accessing resources locally. In addition to improving network performance,

VLANs also provide enhanced security by isolating sensitive data from the other networks and by restricting access to the resources located within the networks.

Virtual SAN (VSAN)

A *virtual SAN/virtual fabric* is a recent evolution of SAN and conceptually, functions in the same way as a VLAN.

In a VSAN, a group of hosts or storage ports communicate with each other using a virtual topology defined on the physical SAN. VSAN technology enables users to build one or more Virtual SANs on a single physical topology containing switches and ISLs. This technology improves storage area network (SAN) scalability, availability, and security. These benefits are derived from the separation of Fibre Channel services in each VSAN and isolation of traffic between VSANs. Some of the features of VSAN are:

- Fibre Channel ID (FC ID) of a host in a VSAN can be assigned to a host in another VSAN, thus improving scalability of SAN.
- Every instance of a VSAN runs all required protocols such as FSPF, domain manager, and zoning.
- Fabric-related configurations in one VSAN do not affect the traffic in another VSAN.
- Events causing traffic disruptions in one VSAN are contained within that VSAN and are not propagated to other VSANs.

5.1.3. Server Virtualization

Server virtualization enables multiple operating systems and applications to run simultaneously on different *virtual machines* created on the same physical server (or group of servers). Virtual machines provide a layer of abstraction between the operating system and the underlying hardware. Within a physical server, any number of virtual servers can be established; depending on hardware capabilities (see Fig. 5.1). Each virtual server seems like a physical machine to the operating system, although all virtual servers share the same underlying physical hardware in an isolated manner. For example, the physical memory is shared between virtual servers but the address space is not. Individual virtual servers can be restarted, upgraded, or even crashed, without affecting the other virtual servers on the same physical machine.

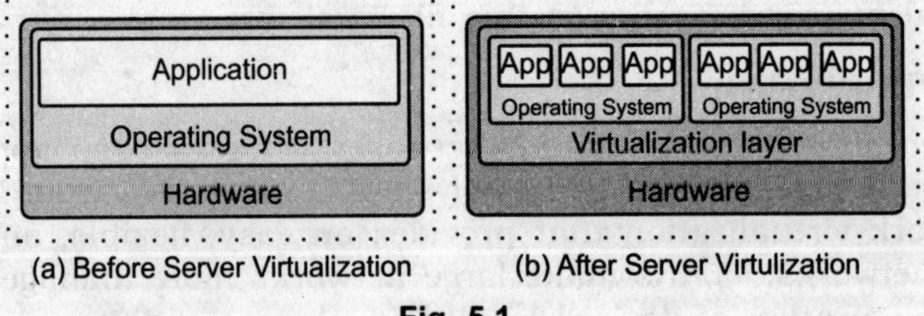

Fig. 5.1.

With changes in computing from a dedicated to a client/server model, the physical server faces resource conflict issues when two or more applications running on these servers have conflicting requirements (e.g., need different values in the same registry entry, different versions of the same DLL). These issues are further compounded with an application's high-availability requirements. As a result, the servers are limited to serve only one application at a time, as shown in Fig. 5.1. (*a*). On the other hand, many applications do not take full advantage of the hardware capabilities available to them. Consequently, resources such as processors, memory, and storage remain underutilized.

Server virtualization addresses the issues that exist in a physical server environment. The virtualization layer, shown in Fig. 5.1. (*b*), helps to overcome resource conflicts by isolating applications running on different operating systems on the same machine. In addition, server virtualization can dynamically move the underutilized hardware resources to a location where they are needed most, improving utilization of the underlying hardware resources.

5.1.4. Storage Virtualization

Storage virtualization is the process of presenting a logical view of the physical storage resources to a host. This logical storage appears and behaves as physical storage directly connected to the host. Throughout the evolution of storage technology, some form of storage virtualization has been implemented. Some examples of storage virtualization are host-based volume management, LUN creation, tape storage virtualization, and disk addressing (CHS to LBA).

The key benefits of storage virtualization include increased storage utilization, adding or deleting storage without affecting an application's availability, and no disruptive data migration (access to files and storage while migrations are in progress).

At the top are four servers, each of which has one virtual volume assigned, which is currently in use by an application. These virtual volumes are mapped to the actual storage in the arrays, as shown at the bottom of the figure. When I/O is sent to a virtual volume, it is redirected through the virtualization at the storage network layer to the mapped physical array.

The discussion that follows provides details about the different types of storage virtualization, methods of implementation, the challenges associated with the implementation of storage virtualization, and examples of implementation.

5.1.5. Application Virtualization

There is also a software angle to "containers" within the cloud. The Web container technology implemented in the cloud greatly impacts developer productivity and flexibility.

The Web container is the part of the application server that manages servlets, Java Server Page (JSP) files, and other Web-tier components. But not all Web container technologies are created equal. Apache Tomcat, for example, is a popular open-source.

Web container technology, but it has limitations for developers who want to go beyond Web-tier applications. If an application needs to use persistence, clustering, failover, messaging, or Enterprise Java Beans (EJBTM), these capabilities have to be added to Tomcat one by one, whereas the Glass Fish

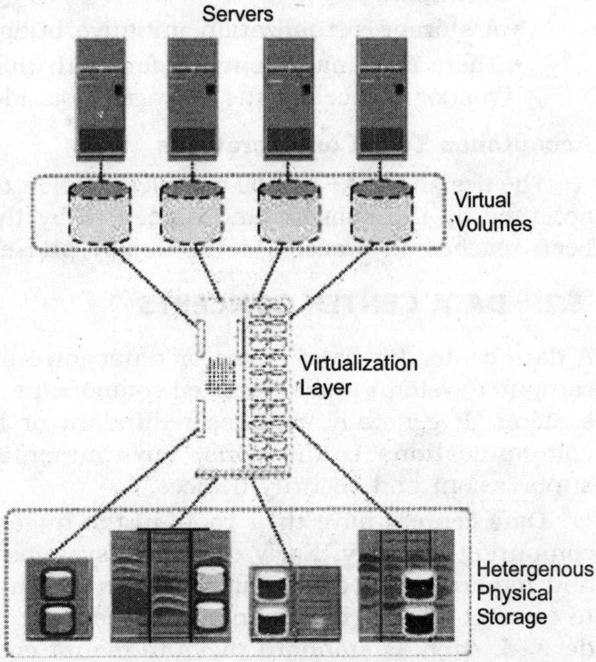

Fig. 5.2.

roject provides an integrated collection of Java EE containers that delivers all of these capabilities.

Today, most cloud computing offerings concentrate on platform virtualization, and the developer chooses the OS and development platform. But increasingly public clouds and certainly private clouds will offer higher-level development environment programming abstractions. Over time, we might expect the level of abstraction that the developer interfaces with to move gradually upward as more functionality percolates down into the platform.

Storage virtualization is now a rapidly maturing technology, but is far from being able to manage heterogeneous arrays of storage in a transparent way. However, the volume based offerings from IBM, HP and Hitachi can make solid contributions to improving utilization and reducing storage management headcount. NAS virtualization is less developed, but homogeneous solutions from NetApp, BlueArc, Hitachi and others are solid. Storage virtualization is an enabling technology for implementing other significant storage improvements, including tiered storage and thin provisioning.

The storage virtualization design and deployment initiative will be implemented when the design systems and arrays are installed, attached and tested, and then approved by the project's key stakeholders.

Expectations (out-of-scope)

The following factors, although necessary for the storage virtualization design and implementation initiative to be successful, are not within the scope of this initiative:

- A project sponsor has been identified with the time and authority to ensure success.
- Storage has to be significantly consolidated, and good common storage management processes and procedures are in place for the storage pools identified.
- A storage virtualization initiative budget has been defined.
- There is management comfort with the trade-off between efficiency and loss of vendor choice for the storage pools identified.

Acceptance Test Considerations

The design phase will be completed when the storage virtualization design has been accepted by the sponsor and agreed to by the key stakeholders, and agreement has been reached to proceed to the deploy phase or kill the initiative.

5.2. DATA CENTER CONCEPTS

A data center (or data centre or datacentre or datacenter) is a facility used to house computer systems and associated components, such as telecommunications and storage systems. It generally includes redundant or backup power supplies, redundant data communications connections, environmental controls (e.g., air conditioning, fire suppression) and security devices.

Data centers have their roots in the huge computer rooms of the early ages of the computing industry. Early computer systems were complex to operate and maintain, and required a special environment in which to operate. Many cables were necessary to connect all the components, and methods to accommodate and organize these were devised, such as standard racks to mount equipment, elevated floors, and cable trays (installed overhead or under the elevated floor). Also, a single mainframe required a great deal of power, and had to be cooled to avoid overheating. Security was important - computers were expensive, and were often used for military purposes. Basic design guidelines for controlling access to the computer room were therefore devised.

During the boom of the microcomputer industry, and especially during the 1980s, computers started to be deployed everywhere, in many cases with little or no care about operating requirements. However, as information technology (IT) operations started to grow in complexity, companies grew aware of the need to control IT resources. With the advent of client-server computing, during the 1990s, microcomputers (now called "servers") started to find their places in the old computer rooms. The availability of inexpensive networking equipment, coupled with new standards for network cabling, made it possible to use a hierarchical design that put the servers in a specific room inside the company. The use of the term "data center", as applied to specially designed computer rooms, started to gain popular recognition about this time.

The boom of data centers came during the dot-com bubble. Companies needed fast Internet connectivity and nonstop operation to deploy systems and establish a presence on the Internet. Installing such equipment was not viable for many smaller companies. Many companies started building very large facilities, called Internet data centers (IDCs), which provide businesses with a range of solutions for systems deployment and operation. New technologies and practices were designed to handle the scale and the operational requirements of such large-scale operations. These practices eventually migrated toward the private data centers, and were adopted largely because of their practical results.

5.3. DATA CENTER REQUIREMENTS

IT operations are a crucial aspect of most organizational operations. One of the main concerns is business continuity; companies rely on their information systems to run their operations. If a system becomes unavailable, company operations may be impaired or stopped completely. It is necessary to provide a reliable infrastructure for IT operations, in order to minimize any chance of disruption. Information security is also a concern, and for this reason a data center has to offer a secure environment which minimizes the chances of a security breach. A data center must therefore keep high standards for assuring the integrity and functionality of its hosted computer environment. This is accomplished through redundancy of both fiber optic cables and power, which includes emergency backup power generation.

Effective data center operation requires a balanced investment in both the facility and the housed equipment. The first step is to establish a baseline facility environment suitable for equipment installation. Standardization and modularity can yield savings and efficiencies in the design and construction of telecommunications data centers.

Standardization means integrated building and equipment engineering. Modularity has the benefits of scalability and easier growth, even when planning forecasts are less than optimal. For these reasons, telecommunications data centers should be planned in repetitive building blocks of equipment, and associated power and support (conditioning) equipment when practical. The use of dedicated centralized systems requires more accurate forecasts of future needs to prevent expensive over construction, or perhaps worse—under construction that fails to meet future needs.

The "lights-out" data center, also known as a darkened or a dark data center, is a data center that, ideally, has all but eliminated the need for direct access by personnel, except under extraordinary circumstances. Because of the lack of need for staff to enter the data center, it can be operated without lighting. All of the devices are accessed and managed by remote systems, with automation programs used to perform unattended operations. In addition to the energy savings, reduction in staffing costs and the ability to locate the site further from population centers, implementing a lights-out data center reduces the threat of malicious attacks upon the infrastructure.

5.3.1. Data Center Classification

The simplest is a Tier 1 data center, which is basically a server room, following basic guidelines for the installation of computer systems. The most stringent level is a Tier 4 data center, which is designed to host mission critical computer systems, with fully redundant subsystems and compartmentalized security zones controlled by biometric access controls methods. Another consideration is the placement of the data center in a subterranean context, for data security as well as environmental considerations such as cooling requirements.

The German Datacenter star audit programe uses an auditing process to certify 5 levels of "gratification" that affect Data Center criticality.

The Uptime Institute, a think tank and professional-services organization based in Santa Fe, New Mexico, defines and holds the copyright on the four levels. The levels describe the availability of data from the hardware at a location. The higher the tier, the greater the availability. The levels are:

Tier Level	Requirements
1.	• Single non-redundant distribution path serving the IT equipment • Non-redundant capacity components • Basic site infrastructure guaranteeing 99.671% availability
2.	• Fulfills all Tier 1 requirements • Redundant site infrastructure capacity components guaranteeing 99.741% availability
3.	• Fulfills all Tier 1 and Tier 2 requirements • Multiple independent distribution paths serving the IT equipment • All IT equipment must be dual-powered and fully compatible with the topology of a site's architecture • Concurrently maintainable site infrastructure guaranteeing 99.982% availability
4.	• Fulfills all Tier 1, Tier 2 and Tier 3 requirements • All cooling equipment is independently dual-powered, including chillers and heating, ventilating and air-conditioning (HVAC) systems • Fault-tolerant site infrastructure with electrical power storage and distribution facilities guaranteeing 99.995% availability

5.3.2. Fire Protection

Data centers feature fire protection systems, including passive and active design elements, as well as implementation of fire prevention programs in operations. Smoke detectors are usually installed to provide early warning of a developing fire by detecting particles generated by smoldering components prior to the development of flame. This allows investigation, interruption of power, and manual fire suppression using hand held fire extinguishers before the fire grows to a large size. A fire sprinkler system is often provided to control a full scale fire if it develops. Fire sprinklers require 18 in (46 cm) of clearance (free of cable trays, etc.) below the sprinklers. Clean agent fire suppression gaseous systems are sometimes installed to suppress a fire earlier than the fire sprinkler system. Passive fire protection elements include the installation of fire walls around the data center, so a fire can be restricted to a portion of the facility for a limited time in the event of the failure of the active fire protection systems, or if they are not installed. For critical facilities these firewalls are often insufficient to protect heat-sensitive electronic equipment, however, because conventional firewall construction is only rated for flame penetration time, not heat penetration. There are also deficiencies in the protection of vulnerable entry points into the server room, such as cable penetrations, coolant line penetrations and air ducts.

5.3.3. Security

Physical security also plays a large role with data centers. Physical access to the site is usually restricted to selected personnel, with controls including bollards and mantraps. Video camera surveillance and permanent security guards are almost always present if the data center is large or contains sensitive information on any of the systems within. The use of finger print recognition man traps is starting to be commonplace.

5.3.4. Energy Use

Energy use is a central issue for data centers. Power draw for data centers ranges from a few kW for a rack of servers in a closet to several tens of MW for large facilities. Some facilities have power densities more than 100 times that of a typical office building. For higher power density facilities, electricity costs are a dominant operating expense and account for over 10% of the total cost of ownership (TCO) of a data center. By 2012 the cost of power for the data center is expected to exceed the cost of the original capital investment.

5.3.5. Energy Efficiency

The most commonly used metric to determine the energy efficiency of a data center is power usage effectiveness, or PUE. This simple ratio is the total power entering the data center divided by the power used by the IT equipment.

$$\text{PUE} = \frac{\text{Total Facility Power}}{\text{IT Equipment Power}}$$

Power used by support equipment, often referred to as overhead load, mainly consists of cooling systems, power delivery, and other facility infrastructure like lighting. The average data center in the US has a PUE of 2.0, meaning that the facility uses one Watt of overhead power for every Watt delivered to IT equipment. State-of-the-art data center energy efficiency is estimated to be roughly 1.2. Some large data center operators like Microsoft and Yahoo! have published projections of PUE for facilities in development.

5.4. DISASTER RECOVERY

Disaster recovery is the process, policies and procedures related to preparing for recovery or continuation of technology infrastructure critical to an organization after a natural or human-induced disaster. Disaster recovery is a subset of business continuity. While business continuity involves planning for keeping all aspects of a business functioning in the midst of disruptive events, disaster recovery focuses on the IT or technology systems that support business functions.

Disaster recovery as a concept developed in the mid to late 1970s as computer center managers began to recognize the dependence of their organizations on their computer systems. At that time most systems were batch-oriented mainframes which in many cases could be down for a number of days before significant damage would be done to the organization.

Backup is a copy of production data, created and retained for the sole purpose of recovering deleted or corrupted data.

With growing business and regulatory demands for data storage, retention, and availability, organizations are faced with the task of backing up an ever-increasing amount of data. This task becomes more challenging as demand for consistent backup and quick restore of data increases throughout the enterprise — which may be spread over multiple sites. Moreover, organizations need to accomplish backup at a lower cost with minimum resources.

Organizations must ensure that the right data is in the right place at the right time. Evaluating backup technologies, recovery, and retention requirements for data and applications is an essential step to ensure successful implementation of the backup and recovery solution. The solution must facilitate easy recovery and retrieval from backups and archives as required by the business.

This chapter includes details about the purposes of backup, strategies for backup and recovery operations, backup methods, the backup architecture, and backup media.

Backups are performed to serve three purposes: disaster recovery, operational backup, and archival.

5.4.1. Disaster Recovery

Backups can be performed to address disaster recovery needs. The backup copies are used for restoring data at an alternate site when the primary site is incapacitated due to a disaster. Based on RPO and RTO requirements, organizations use different backup strategies for disaster recovery. When a tape-based backup method is used as a disaster recovery strategy, the backup tape media is shipped and stored at an offsite location. These tapes can be recalled for restoration at the disaster recovery site. Organizations with stringent RPO and RTO requirements use remote replication technology to replicate data to a disaster recovery site. This allows organizations to bring up production systems online in a relatively short period of time in the event of a disaster.

5.4.2. Operational Backup

Data in the production environment changes with every business transaction and operation. *Operational backup* is a backup of data at a point in time and is used to restore data in the event of data loss or logical corruptions that may occur during routine processing. The majority of restore requests in most organizations fall in this category. For example, it is common for a user to accidentally delete an important e-mail or for a file to become corrupted, which can be restored from operational backup.

Operational backups are created for the active production information by using incremental or differential backup techniques, detailed later in this chapter. An example of an operational backup is a backup performed for a production database just before a bulk batch update. This ensures the availability of a clean copy of the production database if the batch update corrupts the production database.

5.4.3. Archival

Backups are also performed to address archival requirements. Although CAS has emerged as the primary solution for archives, traditional backups are still used by small and medium enterprises for long-term preservation of transaction records, e-mail messages, and other business records required for regulatory compliance.

Apart from addressing disaster recovery, archival, and operational requirements, backups serve as a protection against data loss due to physical damage of a storage device, software failures, or virus attacks. Backups can also be used to protect against accidents such as a deletion or intentional data destruction.

5.5. NETWORK INFRASTRUCTURE

Communications in data centers today are most often based on networks running the IP protocol suite. Data centers contain a set of routers and switches that transport traffic between the servers and to the outside world. Redundancy of the Internet connection is often provided by using two or more upstream service providers.

Some of the servers at the data center are used for running the basic Internet and intranet services needed by internal users in the organization, e.g., e-mail servers, proxy servers, and DNS servers.

Network security elements are also usually deployed: firewalls, VPN gateways, intrusion detection systems, etc. Also common are monitoring systems for the network and some of the applications. Additional off site monitoring systems are also typical, in case of a failure of communications inside the data center.

5.5.1. Applications

Multiple racks of servers, (during server installation) and how a data center commonly looks.

The main purpose of a data center is running the applications that handle the core business and operational data of the organization. Such systems may be proprietary and developed internally by the organization, or bought from enterprise software vendors. Such common applications are ERP and CRM systems.

A data center may be concerned with just operations architecture or it may provide other services as well.

Often these applications will be composed of multiple hosts, each running a single component. Common components of such applications are databases, file servers, application servers, middleware, and various others.

Data centers are also used for off site backups. Companies may subscribe to backup services provided by a data center. This is often used in conjunction with backup tapes. Backups can be taken of servers locally on to tapes. However, tapes stored on site pose a security threat and are also susceptible to fire and flooding. Larger companies may also send their backups off site for added security. This can be done by backing up to a data center. Encrypted backups can be sent over the Internet to another data center where they can be stored securely.

For disaster recovery, several large hardware vendors have developed mobile solutions that can be installed and made operational in very short time. Vendors such as Cisco Systems, Sun Microsystems, Bull IBM, and HP have developed systems that could be used for this purpose.

5.6. PRINCIPLES MANAGING AND MONITORING

As IT systems have become increasingly critical to the smooth operation of a company, and arguably the economy as a whole, the importance of ensuring the continued operation of those systems, or the rapid recovery of the systems, has increased.

It is estimated that most large companies spend between 2% and 4% of their IT budget on disaster recovery planning, with the aim of avoiding larger losses in the event that the business cannot continue to function due to loss of IT infrastructure and data. Of companies that had a major loss of business data, 43% never reopen, 51% close within two years, and only 6% will survive long-term.

As a result, preparation for continuation or recovery of systems needs to be taken very seriously. This involves a significant investment of time and money with the aim of ensuring minimal losses in the event of a disruptive event.

5.6.1. Classification of Disasters

Disasters can be classified in two broad categories. The first is natural disasters such as floods, hurricanes, tornadoes or earthquakes. While preventing a natural disaster is very difficult, measures such as good planning which includes mitigation measures can help reduce or avoid losses. The second category is man made disasters. These include hazardous material spills, infrastructure failure, or bio-terrorism. In these instances surveillance and mitigation planning are invaluable towards avoiding or lessening losses from these events.

5.6.2. Control Measures in Recovery Plan

Control measures are steps or mechanisms that can reduce or eliminate various threats for organizations. Different types of measures can be included in BCP/DRP.

Disaster recovery planning is a subset of a larger process known as business continuity planning and should include planning for resumption of applications, data, hardware, communications (such as networking) and other IT infrastructure. A business continuity plan (BCP) includes planning for non-IT related aspects such as key personnel, facilities, crisis communication and reputation protection, and should refer to the disaster recovery plan (DRP) for IT related infrastructure recovery/continuity. This article focuses on disaster recovery planning as related to IT infrastructure. Types of measures:

1. **Preventive measures:** These controls are aimed at preventing an event from occurring.
2. **Detective measures:** These controls are aimed at detecting or discovering unwanted events.
3. **Corrective measures:** These controls are aimed at correcting or restoring the system after disaster or event.

These controls should be always documented and tested regularly.

5.6.3. Strategies

Prior to selecting a disaster recovery strategy, a disaster recovery planner should refer to their organization's business continuity plan which should indicate the key metrics of recovery point objective (RPO) and recovery time objective (RTO) for various business processes (such as the process to run payroll, generate an order, etc.). The metrics specified for the business processes must then be mapped to the underlying IT systems and infrastructure that support those processes.

Once the RTO and RPO metrics have been mapped to IT infrastructure, the DR planner can determine the most suitable recovery strategy for each system. An important note here however is that the business ultimately sets the IT budget and therefore the RTO and RPO metrics need to fit with the available budget. While most business unit heads would like zero data loss and zero time loss, the cost associated with that level of protection may make the desired high availability solutions impractical.

The following is a list of the most common strategies for data protection.

- Backups made to tape and sent off-site at regular intervals.
- Backups made to disk on-site and automatically copied to off-site disk, or made directly to off-site disk.
- Replication of data to an off-site location, which overcomes the need to restore the data (only the systems then need to be restored or synchronized). This generally makes use of storage area network (SAN) technology.
- High availability systems which keep both the data and system replicated off-site, enabling continuous access to systems and data.

In many cases, an organization may elect to use an outsourced disaster recovery provider to provide a stand-by site and systems rather than using their own remote facilities.

In addition to preparing for the need to recover systems, organizations must also implement precautionary measures with an objective of preventing a disaster in the first place. These may include some of the following:

- Local mirrors of systems and/or data and use of disk protection technology such as RAID.
- Surge protectors - to minimize the effect of power surges on delicate electronic equipment.
- Uninterruptible power supply (UPS) and/or backup generator to keep systems going in the event of a power failure.
- Fire preventions - alarms, fire extinguishers.
- Anti-virus software and other security measures.

5.7. INFORMATION SECURITY

Information Security Components: or qualities, *i.e.*, Confidentiality, Integrity and Availability (CIA). Information Systems are decomposed in three main portions, hardware, software and communications with the purpose to identify and apply information security industry standards, as mechanisms of protection and prevention, at three levels or layers: physical, personal and organizational. Essentially, procedures or policies are implemented to tell people (administrators, users and operators)how to use products to ensure information security within the organizations.

Information security means protecting information and information systems from unauthorized access, use, disclosure, disruption, modification, perusal, inspection, recording or destruction.

The terms information security, computer security and information assurance are frequently incorrectly used interchangeably. These fields are interrelated often and share the common goals of protecting the confidentiality, integrity and availability of information; however, there are some subtle differences between them.

These differences lie primarily in the approach to the subject, the methodologies used, and the areas of concentration. Information security is concerned with the confidentiality, integrity and availability of data regardless of the form the data may take: electronic, print, or other forms.

Computer security can focus on ensuring the availability and correct operation of a computer system without concern for the information stored or processed by the computer.

Governments, military, corporations, financial institutions, hospitals and private businesses amass a great deal of confidential information about their employees, customers, products, research, and financial status. Most of this information is now collected, processed and stored on electronic computers and transmitted across networks to other computers.

Should confidential information about a business' customers or finances or new product line fall into the hands of a competitor, such a breach of security could lead to lost business, law suits or even bankruptcy of the business. Protecting confidential information is a business requirement, and in many cases also an ethical and legal requirement.

For the individual, information security has a significant effect on privacy, which is viewed very differently in different cultures.

The field of information security has grown and evolved significantly in recent years. There are many ways of gaining entry into the field as a career. It offers many areas for specialization including: securing network(s) and allied infrastructure, securing applications and databases, security testing, information systems auditing, business continuity planning and digital forensics science, etc.

This article presents a general overview of information security and its core concepts.

Since the early days of writing, heads of state and military commanders understood that it was necessary to provide some mechanism to protect the confidentiality of written correspondence and to have some means of detecting tampering.

Julius Caesar is credited with the invention of the Caesar cipher ca. 50 B.C., which was created in order to prevent his secret messages from being read should a message fall into the wrong hands.

World War II brought about many advancements in information security and marked the beginning of the professional field of information security.

The end of the 20th century and early years of the 21st century saw rapid advancements in telecommunications, computing hardware and software, and data encryption. The availability of smaller, more powerful and less expensive computing equipment made electronic data processing within the reach of small business and the home user. These computers quickly became interconnected through a network generically called the Internet or World Wide Web.

The rapid growth and widespread use of electronic data processing and electronic business conducted through the Internet, along with numerous occurrences of international terrorism, fueled the need for better methods of protecting the computers and the information they store, process and transmit. The academic disciplines of computer security, information security and information assurance emerged along with numerous professional organizations—all sharing the common goals of ensuring the security and reliability of information systems.

5.7.1. Basic Principles

For over twenty years, information security has held confidentiality, integrity and availability (known as the CIA triad) to be the core principles of information security.

There is continuous debate about extending this classic trio. Other principles such as Accountability have sometimes been proposed for addition–it has been pointed out that issues such as Non-Repudiation do not fit well within the three core concepts, and as regulation of computer systems has increased (particularly amongst the Western nations) Legality is becoming a key consideration for practical security installations.

5.7.2. Confidentiality

Confidentiality is the term used to prevent the disclosure of information to unauthorized individuals or systems. For example, a credit card transaction on the Internet requires the credit card number to be transmitted from the buyer to the merchant and from the merchant to a transaction processing network. The system attempts to enforce confidentiality by encrypting the card number during transmission, by limiting the places where it might appear (in databases, log files, backups, printed receipts, and so on), and by restricting access to the places where it is stored. If an unauthorized party obtains the card number in any way, a breach of confidentiality has occurred.

Breaches of confidentiality take many forms. Permitting someone to look over your shoulder at your computer screen while you have confidential data displayed on it could be a breach of confidentiality. If a laptop computer containing sensitive information about a company's employees is stolen or sold, it could result in a breach of confidentiality. Giving out confidential information over the telephone is a breach of confidentiality if the caller is not authorized to have the information.

Confidentiality is necessary (but not sufficient) for maintaining the privacy of the people whose personal information a system holds.[citation needed]

5.7.3. Integrity

In information security, integrity means that data cannot be modified undetectably.[citation needed] This is not the same thing as referential integrity in databases, although it can be viewed as a special case of Consistency as understood in the classic ACID model of transaction processing. Integrity is violated when a message is actively modified in transit. Information security systems typically provide message integrity in addition to data confidentiality.

5.7.4. Availability

For any information system to serve its purpose, the information must be available when it is needed. This means that the computing systems used to store and process the information, the security controls used to protect it, and the communication channels used to access it must be functioning correctly. High availability systems aim to remain available at all times, preventing service disruptions due to power outages, hardware failures, and system upgrades. Ensuring availability also involves preventing denial-of-service attacks.

5.7.5. Authenticity

In computing, e-Business and information security it is necessary to ensure that the data, transactions, communications or documents (electronic or physical) are genuine.

It is also important for authenticity to validate that both parties involved are who they claim they are.

5.7.6. Non-Repudiation

In law, non-repudiation implies one's intention to fulfill their obligations to a contract. It also implies that one party of a transaction cannot deny having received a transaction nor can the other party deny having sent a transaction.

Electronic commerce uses technology such as digital signatures and public key encryption to establish authenticity and non-repudiation.

5.7.7. Risk Management

A comprehensive treatment of the topic of risk management is beyond the scope of this article. However, a useful definition of risk management will be provided as well as some basic terminology and a commonly used process for risk management.

There are two things in this definition that may need some clarification. First, the process of risk management is an ongoing iterative process. It must be repeated indefinitely. The business environment is constantly changing and new threats and vulnerability emerge every day. Second, the choice of countermeasure (computer)s (controls) used to manage risks must strike a balance between productivity, cost, effectiveness of the countermeasure, and the value of the informational asset being protected.

Risk is the likelihood that something bad will happen that causes harm to an informational asset (or the loss of the asset). A vulnerability is a weakness that could be used to endanger or cause harm to an informational asset. A threat is anything (man made or act of nature) that has the potential to cause harm.

The likelihood that a threat will use a vulnerability to cause harm creates a risk. When a threat does use a vulnerability to inflict harm, it has an impact. In the context of information security, the impact is a loss of availability, integrity, and confidentiality, and possibly other losses (lost income, loss of life, loss of real property). It should be pointed out that it is not possible to identify all risks, nor is it possible to eliminate all risk. The remaining risk is called *residual risk.*

A risk assessment is carried out by a team of people who have knowledge of specific areas of the business. Membership of the team may vary over time as different parts of the business are assessed. The assessment may use a subjective qualitative analysis based on informed opinion, or where reliable dollar figures and historical information is available, the analysis may use quantitative analysis.

In broad terms, the risk management process consists of:

1. **Identification of assets and estimating their value. Include:** people, buildings, hardware, software, data (electronic, print, other), supplies.
2. **Conduct a threat assessment. Include:** Acts of nature, **acts of war**, accidents, malicious acts originating from inside or outside the organization.
3. Conduct a **vulnerability assessment**, and for each vulnerability, calculate the probability that it will be exploited. Evaluate policies, procedures, standards, training, **physical security, quality control**, technical security.
4. Calculate the impact that each threat would have on each asset. Use qualitative analysis or quantitative analysis.
5. Identify, select and implement appropriate controls. Provide a proportional response. Consider productivity, cost effectiveness, and value of the asset.

6. Evaluate the effectiveness of the control measures. Ensure the controls provide the required cost effective protection without discernible loss of productivity.

For any given risk, Executive Management can choose to accept the risk based upon the relative low value of the asset, the relative low frequency of occurrence, and the relative low impact on the business. Or, leadership may choose to mitigate the risk by selecting and implementing appropriate control measures to reduce the risk. In some cases, the risk can be transferred to another business by buying insurance or out-sourcing to another business. The reality of some risks may be disputed. In such cases leadership may choose to deny the risk. This is itself a potential risk.

5.7.8. Controls

When Management chooses to mitigate a risk, they will do so by implementing one or more of three different types of controls.

Administrative controls (also called procedural controls) consist of approved written policies, procedures, standards and guidelines. Administrative controls form the framework for running the business and managing people. They inform people on how the business is to be run and how day-to-day operations are to be conducted. Laws and regulations created by government bodies are also a type of administrative control because they inform the business. Some industry sectors have policies, procedures, standards and guidelines that must be followed—the Payment Card Industry (PCI) Data Security Standard required by Visa and Master Card is such an example. Other examples of administrative controls include the corporate security policy, password policy, hiring policies, and disciplinary policies.

Administrative controls form the basis for the selection and implementation of logical and physical controls. Logical and physical controls are manifestations of administrative controls. Administrative controls are of paramount importance.

5.7.9. Logical

Logical controls (also called technical controls) use software and data to monitor and control access to information and computing systems. For example: passwords, network and host based firewalls, network intrusion detection systems, access control lists, and data encryption are logical controls.

An important logical control that is frequently overlooked is the principle of least privilege. The principle of least privilege requires that an individual, program or system process is not granted any more access privileges than are necessary to perform the task. A blatant example of the failure to adhere to the principle of least privilege is logging into Windows as user Administrator to read Email and surf the Web. Violations of this principle can also occur when an individual collects additional access privileges over time. This happens when employees' job duties change, or they are promoted to a new position, or they transfer to another department. The access privileges required by their new duties are frequently added onto their already existing access privileges which may no longer be necessary or appropriate.

5.7.10. Physical

Physical controls monitor and control the environment of the work place and computing facilities. They also monitor and control access to and from such facilities. For example: doors, locks, heating and air conditioning, smoke and fire alarms, fire suppression systems, cameras, barricades, fencing, security guards, cable locks, etc. Separating the network and work place into functional areas are also physical controls.

An important physical control that is frequently overlooked is the separation of duties. Separation of duties ensures that an individual can not complete a critical task by himself. For example: an employee who submits a request for reimbursement should not also be able to authorize payment or print the check. An applications programmer should not also be the server administrator or the database administrator - these roles and responsibilities must be separated from one another.

5.8. SECURITY CLASSIFICATION FOR INFORMATION

An important aspect of information security and risk management is recognizing the value of information and defining appropriate procedures and protection requirements for the information. Not all information is equal and so not all information requires the same degree of protection. This requires information to be assigned a security classification.

The first step in information classification is to identify a member of senior management as the owner of the particular information to be classified. Next, develop a classification policy. The policy should describe the different classification labels, define the criteria for information to be assigned a particular label, and list the required security controls for each classification.

Some factors that influence which classification information should be assigned include how much value that information has to the organization, how old the information is and whether or not the information has become obsolete. Laws and other regulatory requirements are also important considerations when classifying information.

The type of information security classification labels selected and used will depend on the nature of the organisation, with examples being:

- **In the business sector, labels such as:** Public, Sensitive, Private, Confidential.
- **In the government sector, labels such as:** Unclassified, Sensitive But Unclassified, Restricted, Confidential, Secret, Top Secret and their non-English equivalents.
- **In cross-sectoral formations, the Traffic Light Protocol, which consists of:** White, Green, Amber and Red.

All employees in the organization, as well as business partners, must be trained on the classification schema and understand the required security controls and handling procedures for each classification. The classification of a particular information asset has been assigned should be reviewed periodically to ensure the classification is still appropriate for the information and to ensure the security controls required by the classification are in place.

5.8.1. Access Control

Access to protected information must be restricted to people who are authorized to access the information. The computer programs, and in many cases the computers that process the information, must also be authorized. This requires that mechanisms be in place to control the access to protected information. The sophistication of the access control mechanisms should be in parity with the value of the information being protected—the more sensitive or valuable the information the stronger the control mechanisms need to be. The foundation on which access control mechanisms are built start with identification and authentication.

Identification is an assertion of who someone is or what something is. If a person makes the statement "*Hello, my name is John Doe*" they are making a claim of who they are. However, their claim may or may not be true. Before John Doe can be granted access to protected information it will be necessary to verify that the person claiming to be John Doe really is John Doe.

Authentication is the act of verifying a claim of identity. When John Doe goes into a bank to make a withdrawal, he tells the bank teller he is John Doe (a claim of identity). The bank teller asks to see a photo ID, so he hands the teller his driver's license. The bank teller checks the license to make sure it has John Doe printed on it and compares the photograph on the license against the person claiming to be John Doe. If the photo and name match the person, then the teller has authenticated that John Doe is who he claimed to be.

There are three different types of information that can be used for authentication: something you know, something you have, or something you are. Examples of something you know include such things as a PIN, a password, or your mother's maiden name. Examples of *something you have* include a driver's license or a magnetic swipe card. *Something* you are refers to biometrics. Examples of biometrics include palm prints, finger prints, voice prints and retina (eye) scans. Strong authentication requires providing information from two of the three different types of authentication information. For example, something you know plus something you have. This is called two factor authentication.

On computer systems in use today, the username is the most common form of identification and the password is the most common form of authentication. Usernames and passwords have served their purpose but in our modern world they are no longer adequate. Usernames and passwords are slowly being replaced with more sophisticated authentication mechanisms.

After a person, program or computer has successfully been identified and authenticated then it must be determined what informational resources they are permitted to access and what actions they will be allowed to perform (run, view, create, delete, or change). This is called authorization.

Authorization to access information and other computing services begins with administrative policies and procedures. The policies prescribe what information and computing services can be accessed, by whom, and under what conditions. The access control mechanisms are then configured to enforce these policies.

Different computing systems are equipped with different kinds of access control mechanisms—some may even offer a choice of different access control mechanisms. The access control mechanism a system offers will be based upon one of three approaches to access control or it may be derived from a combination of the three approaches.

The non-discretionary approach consolidates all access control under a centralized administration. The access to information and other resources is usually based on the individuals function (role) in the organization or the tasks the individual must perform. The discretionary approach gives the creator or owner of the information resource the ability to control access to those resources. In the Mandatory access control approach, access is granted or denied basing upon the security classification assigned to the information resource.

Examples of common access control mechanisms in use today include Role-based access control available in many advanced Database Management Systems, simple file permissions provided in the UNIX and Windows operating systems, Group Policy Objects provided in Windows network systems, Kerberos, RADIUS, TACACS and the simple access lists used in many firewalls and routers.

To be effective, policies and other security controls must be enforceable and upheld. Effective policies ensure that people are held accountable for their actions. All failed and successful authentication attempts must be logged, and all access to information must leave some type of audit trail.

5.8.2. Cryptography

Information security uses cryptography to transform usable information into a form that renders it unusable by anyone other than an authorized user; this process is called encryption. Information that has been encrypted (rendered unusable) can be transformed back into its original usable form by an authorized user, who possesses the cryptographic key, through the process of decryption. Cryptography is used in information security to protect information from unauthorized or accidental disclosure while the information is in transit (either electronically or physically) and while information is in storage.

Cryptography provides information security with other useful applications as well including improved authentication methods, message digests, digital signatures, non-repudiation, and encrypted network communications. Older less secure application such as telnet and ftp are slowly being replaced with more secure applications such as ssh that use encrypted network communications. Wireless communications can be encrypted using protocols such as WPA/WPA2 or the older (and less secure) WEP. Wired communications (such as ITU-T G.hn) are secured using AES for encryption and X.1035 for authentication and key exchange. Software applications such as GnuPG or PGP can be used to encrypt data files and Email.

Cryptography can introduce security problems when it is not implemented correctly. Cryptographic solutions need to be implemented using industry accepted solutions that have undergone rigorous peer review by independent experts in cryptography. The length and strength of the encryption key is also an important consideration. A key that is weak or too short will produce weak encryption. The keys used for encryption and decryption must be protected with the same degree of rigor as any other confidential information. They must be protected from unauthorized disclosure and destruction and they must be available when needed. PKI solutions address many of the problems that surround key management.

5.8.3. Process

The terms reasonable and prudent person, due care and due diligence have been used in the fields of Finance, Securities, and Law for many years. In recent years these terms have found their way into the fields of computing and information security. U.S.A. Federal Sentencing Guidelines now make it possible to hold corporate officers liable for failing to exercise due care and due diligence in the management of their information systems.

In the business world, stockholders, customers, business partners and governments have the expectation that corporate officers will run the business in accordance with accepted business practices and in compliance with laws and other regulatory requirements. This is often described as the "reasonable and prudent person" rule. A prudent person takes due care to ensure that everything necessary is done to operate the business by sound business principles and in a legal ethical manner. A prudent person is also diligent (mindful, attentive, and ongoing) in their due care of the business.

In the field of Information Security, Harris offers the following definitions of due care and due diligence:

"*Due care are steps that are taken to show that a company has taken responsibility for the activities that take place within the corporation and has taken the necessary steps to help protect the company, its resources, and employees.*" And, [Due diligence] are the "*continual activities that make sure the protection mechanisms are continually maintained and operational.*"

Attention should be made to two important points in these definitions. First, in due care, steps are taken to show—this means that the steps can be verified, measured, or even produce tangible artifacts. Second, in due diligence, there are continual activities—this means that people are actually doing things to monitor and maintain the protection mechanisms, and these activities are ongoing.

5.9. INDUSTRY MANAGEMENT STANDARDS (SNMP, SMI-S, CIM)

Until the adoption of SMI-S, management of a storage area network (SAN) used either the Simple Network Management Protocol (SNMP) or swapping of vendor specific Application Programming Interfaces (API). Neither meet current needs.

The Storage Networking Industry Association (SNIA) has been engaged in an initiative to develop a common, open storage, and SAN management interface. SMI-S is based on Web-Based Enterprise Management (WBEM) technology and the DMTF's Common Information Model (CIM). The initiative was formally created to enable broad interoperability among heterogeneous storage vendor systems and to enable better management solutions that span these environments.

This initiative is known as the Storage Management Initiative (SMI). For more information, see www.snia.org. The SMI Specification, known as SMI-S, offers substantial benefits to users and vendors. It forms a normalized, abstracted model to which a storage infrastructure's physical and logical components can be mapped, and which can be used by management applications such as storage resource management, device management, and data management for standardized, effective, end-to-end control of storage resources (see Fig. 5.3).

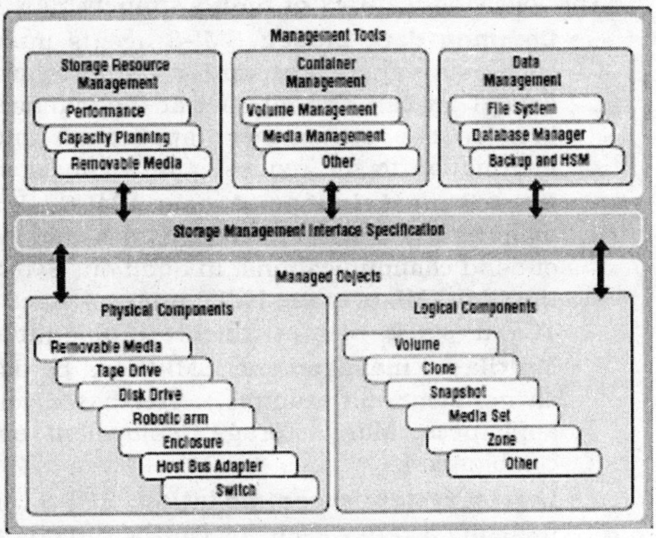

Fig. 5.3.

Using SMI-S, the storage software developers have a single normalized and unified object model comprising the detailed document that contains information about managing the breadth of SAN components. Moreover, SMI-S eliminates the need for development of vendor-proprietary management interfaces, enabling vendors to focus on added value functions and offering solutions in a way that will support new devices as long as they adhere to the standard. Using SMI-S, device vendors can build new features and functions to manage storage subsystems and expose them via SMI-S. The SMI-S-compliant products lead to easier, faster deployment, and accelerated adoption of policy-based storage management frameworks.

The information required to perform management tasks is better organized or structured in a way that enables disparate groups of people to use it. This can be accomplished by developing a model or representation of the details required by users working within a particular domain. Such an approach is referred to as an *information model.* An information model requires a set of legal statements or syntax to capture the representation and expressions necessary to manage common aspects of that domain.

The CIM is a language and methodology for describing management elements. A CIM schema includes models for systems, applications, networks, and devices. This schema also enables applications from different vendors working on different platforms to describe the management data in a standard format so that it can be shared among a variety of management applications.

WBEM is an initiative in the DMTF. It is a set of technologies that enables interoperable management of an enterprise. WBEM consists of CIM, an XML DTD defining the tags (XML encodings) to describe the CIM Schema and its data, and a set of HTTP operations for exchanging the XML-based information.

The main objective of developing a WBEM is to unify the management of enterprise-computing environments that were traditionally administered through the management stacks, such as SNMP and Common Management Information Protocol (CMIP).

WBEM includes the CIM standard, the xmlCIM encoding specification, and a transport mechanism — CIM operation over HTTP.

The following features of SMI-S simplify SAN management:

- **Common data model:** SMI-S agents interact with an SMI-S-enabled device, such as a switch, a server, or a storage array, to extract relevant management data. They can also interact at the management layer to exchange information between one management application and another. They then provide this information to the requester in a consistent syntax and format.
- **Interconnect independence:** SMI-S eliminates the need to redesign the management transport and enables components to be managed by using out-of-band communications. In addition, SMI-S offers the advantages of specifying the CMI-XML over the HTTP protocol stack and utilizing the lower layers of the TCP/IP stack, both of which are ubiquitous in today's networking world.
- **Multilayer management:** SMI-S can be used in a multilayer and crossdomain environment—for example, server-based volume managers and network storage appliances. Many storage deployment environments currently employ this combination.
- **Legacy system accommodation:** SMI-S can be used to manage legacy systems by using a proxy agent or can be directly supported by the device itself. SMI-S can coexist with proprietary APIs and agents as well as providing integration framework for such mechanisms.
- **Policy-based management:** SMI-S includes object models applicable across all classes of devices, enabling a SAN administrator to implement policy-based management for entire storage networks.

A common information model eases the huge tasks of interpreting relations and correlations of information across technology stacks and vendor devices. A CIM makes it easier for management applications as well as device/solution developers, as it creates normalized abstracted targets when both management and managed elements interact.

SNMP

The most common system management infrastructure is based upon SNMP. Its beginnings go back to the 1980's. Its original charter was to provide a common framework for managing the low level components of a computer network such as routers and bridges. Over time, SNMP became the accepted standard for discovering network topology and monitoring its status.

However, network technology has evolved. Enterprises have increasingly integrated computers into the operation of their business. Storage resources have become more distributed. Such trends impose greater demands on a company to efficiently manage its enterprise computing environment. Unfortunately, despite attempts to expand its capabilities, SNMP can no longer fulfill these demands. The reasons are many.

SNMP definitions and data are represented in a text file called a Management Information Base (MIB). Modeling a new object in SNMP requires defining a new MIB. One MIB has no inherent relationship to any other MIB. A separate MIB is needed for each defined SNMP managed object. Achieving industry interoperability requires that each participating vendor agrees to use the same MIBs in the same way.

One MIB has no inherent relationship to any other MIB. Although one MIB may contain some of the same properties as another MIB, no SNMP mechanism exists to take advantage of the commonality. Furthermore, SNMP cannot easily model objects in an enterprise or storage area network (SAN). SNMP was originally chartered to only manage low level network components, not the entire enterprise. Its naming scheme and infrastructure do not lend themselves to management of diverse enterprise environments. For example, SNMP cannot model non-computing elements such as corporate organizational units.

SNMP defines only four operations. They are *Get, GetNext, Set* and *Trap.* Get retrieves one or more MIB values. GetNext is used to sequentially retrieve data from a table of MIB values, one row at a time. A table is a logical structuring of MIB values into rows and columns. *Set* is used to update an individual MIB value. However, almost all SNMP management frameworks are read-only while SANs require active management. For example, an storage administrator must have the ability to create new storage capacity or to reconfigure existing storage capacity.

For these reasons, the Storage Management Initiative (SMI) chose the Common Information Model (CIM) because CIM is a much more extensible data model. The SMI chose Web Based Enterprise Management (WBEM) as the underlying infrastructure technology because it supplied a richer set of capabilities.

API Swapping

Current enterprise networks consist of systems, network and storage resources from many different vendors. Each vendor created their own private management application that used a private interface. Remote management even used private network protocols. Consequently, in the worst case scenario, a storage management administrator has to use a separate management application for each storage product.

One solution is create a management application that is knowledgeable of more than one private interface or protocol. However, creating such an application requires vendors to share information about their interfaces. Such sharing is called API swapping.

Unfortunately, API swapping does not scale business-wise or engineering-wise. First, the two competitors must negotiate a mutually beneficial business arrangement. Next, the two competitors must integrate the new technology into each other's management applications. This time-intensive process must be repeated for each new device to be supported from a new vendor. Additionally, even after the initial integration, continual updates are required for each individual product release.

The SMI chose CIM and WBEM as the underlying technology because they were based on open standards and provided a normalized approach to the management of SAN resources. Using CIM and WBEM allows storage management application vendors and storage device vendors to interact in a common manner without knowledge of any private API's.

The Common Information Model (CIM) is an open standard that defines how managed elements in an IT environment are represented as a common set of objects and relationships between them. This is intended to allow consistent management of these managed elements, independent of their manufacturer or provider.

One way to describe CIM is to say that it allows multiple parties to exchange management information about these managed elements. However, this falls short in expressing that CIM not only represents these managed elements and the management information, but also provides means to actively control and manage these elements. By using a common model of information, management software can be written once and work with many implementations of the common model without complex and costly conversion operations or loss of information.

The CIM standard is defined and published by the Distributed Management Task Force (DMTF). A related standard is Web-Based Enterprise Management (WBEM, also defined by DMTF) which defines a particular implementation of CIM, including protocols for discovering and accessing such CIM implementations.

Schema and Specifications

The CIM standard includes the CIM Infrastructure Specification and the CIM Schema:

- **CIM Infrastructure Specification:** The CIM Infrastructure Specification defines the architecture and concepts of CIM, including a language by which the CIM Schema (including any extension schema) is defined, and a method for mapping CIM to other information models, such as SNMP. The CIM architecture is based upon UML, so it is object-oriented: The managed elements are represented as CIM classes and any relationships between them are represented as CIM associations. Inheritance allows specialization of common base elements into more specific derived elements.
- **CIM Schema:** The CIM Schema is a conceptual schema which defines the specific set of objects and relationships between them that represent a common base for the managed elements in an IT environment. The CIM Schema covers most of today's elements in an IT environment, for example computer systems, operating systems, networks, middleware, services and storage. The CIM Schema defines a common basis for representing these managed elements. Since most managed elements have product and vendor specific behavior, the CIM Schema is extensible in order to allow the producers of these elements to represent their specific features seamlessly together with the common base functionality defined in the CIM Schema.

CIM is the basis for most of the other DMTF standards (e.g. WBEM or SMASH). It is also the basis for the SMI-S standard for storage management

Storage Management Initiative—Specification

SMI-S, or the Storage Management Initiative: Specification, is a storage standard developed and maintained by the Storage Networking Industry Association (SNIA). It has also been ratified as an ISO standard. SMI-S is based upon the Common Information Model and the Web-Based Enterprise Management standards defined by the Distributed Management Task Force, which define management functionality via HTTP. The most recent approved version of SMI-S is available at the SNIA.

The main objective of SMI-S is to enable broad interoperable management of heterogeneous storage vendor systems. The current version is SMI-S V1.5.0. Over 75 software products and over 800 hardware products are certified as conformant to SMI-S.

Basic Concepts

SMI-S defines CIM management profiles for storage systems. The complete SMI Specification is categorized in profiles and subprofiles. A profile describes the behavioral aspects of an autonomous, self-contained management domain. SMI-S includes profiles for Arrays, Switches, Storage Virtualizers, Volume Management and many other domains. In DMTF parlance, a provider is an implementation for a specific profile. A subprofile describes part of the domain, which can be a common part in many profiles.

At a very basic level, SMI-S entities are divided into two categories:

- **Clients** are management software applications that can reside virtually anywhere within a network, provided they have a communications link (either within the data path or outside the data path) to providers.
- **Servers** are the devices under management. Servers can be disk arrays, virtualization engines, host bus adapters, switches, tape drives, etc.

5.10. STANDARD FRAMEWORK APPLICATIONS

Application Framework

In computer programming, an application framework consists of a software framework used by software developers to implement the standard structure of an application for a specific development environment (such as an operating system or a web application).

Application frameworks became popular with the rise of graphical user interfaces (GUIs), since these tended to promote a standard structure for applications. Programmers find it much simpler to create automatic GUI creation tools when using a standard framework, since this defines the underlying code structure of the application in advance. Developers usually use object-oriented programming techniques to implement frameworks such that the unique parts of an application can simply inherit from pre-existing classes in the framework.

Examples

Apple Computer developed one of the first commercial application frameworks, MacApp (first released in 1985), for the Macintosh. Originally written in an extended (object-oriented) version of Pascal, it later appeared rewritten in C++. Other popular frameworks for the Mac include Metrowerks' PowerPlant and MacZoop (both based on Carbon). Cocoa for Mac OS X offers a different approach to an application framework, one based upon the OPENSTEP framework developed at NeXT.

Free-software frameworks exist as part of the Mozilla, OpenOffice.org, GNOME, KDE, and Eclipse projects.

Microsoft markets a framework for developing Windows applications in C++ called the Microsoft Foundation Class Library.

A number of frameworks can build cross-platform applications for Linux, Macintosh, and Windows from the same source code, such as Qt, the widget toolkits wxWidgets, FOX toolkit, or Eclipse RCP.

Oracle Application Development Framework (Oracle ADF) aids in producing Java-oriented systems.

SUMMARY

1. Storage virtualization is now a rapidly maturing technology, but is far from being able to manage heterogeneous arrays of storage in a transparent way. However, the volume based offerings from IBM, HP and Hitachi can make solid contributions to improving utilization and reducing storage management headcount.
2. NAS virtualization is less developed, but homogeneous solutions from NetApp, BlueArc, Hitachi and others are solid. Storage virtualization is an enabling technology for implementing other significant storage improvements, including tiered storage and thin provisioning.
3. Storage has be significantly consolidated, and good common storage management processes and procedures are in place for the storage pools identified.
4. A storage virtualization initiative budget has been defined. There is management comfort with the trade-off between efficiency and loss of vendor choice for the storage pools identified.
5. In NAS environments, homogeneous solutions for directory integration will significantly lower the project risk. There would need to be compelling functionality and additional testing before large-scale heterogeneous NAS virtualization solutions are implemented.
6. A data center (or data centre or datacentre or datacenter) is a facility used to house computer systems and associated components, such as telecommunications and storage systems. It generally includes redundant or backup power supplies, redundant data communications connections, environmental controls (e.g., air conditioning, fire suppression) and security devices.
7. Information security is also a concern, and for this reason a data center has to offer a secure environment which minimizes the chances of a security breach. A data center must therefore keep high standards for assuring the integrity and functionality of its hosted computer environment. This is accomplished through redundancy of both fiber optic cables and power, which includes emergency backup power generation.
8. IT operations are a crucial aspect of most organizational operations. One of the main concerns is business continuity; companies rely on their information systems to run their operations. If a system becomes unavailable, company operations may be impaired or stopped completely. It is necessary to provide a reliable infrastructure for IT operations, in order to minimize any chance of disruption.
9. Effective data center operation requires a balanced investment in both the facility and the housed equipment. The first step is to establish a baseline facility environment suitable for equipment installation. Standardization and modularity can yield savings and efficiencies in the design and construction of telecommunications data centers.
10. Standardization means integrated building and equipment engineering. Modularity has the benefits of scalability and easier growth, even when planning forecasts are less than optimal. For these reasons, telecommunications data centers should be planned in repetitive building blocks of equipment, and associated power and support (conditioning) equipment when practical.

QUESTION WITH ANSWER

Q.1. What is the Virtualization of Memory?

Ans. Storage virtualization is now a rapidly maturing technology, but is far from being able to manage heterogeneous arrays of storage in a transparent way. However, the volume based offerings from IBM, HP and Hitachi can make solid contributions to improving utilization and reducing storage management headcount. NAS virtualization is less developed, but homogeneous solutions from NetApp, BlueArc, Hitachi and others are solid. Storage virtualization is an enabling technology for implementing other significant storage improvements, including tiered storage and thin provisioning.

The storage virtualization design and deployment initiative will be implemented when the design systems and arrays are installed, attached and tested, and then approved by the project's key stakeholders.

Expectations (out-of-scope): The following factors, although necessary for the storage virtualization design and implementation initiative to be successful, are not within the scope of this initiative:

- A project sponsor has been identified with the time and authority to ensure success.
- Storage has be significantly consolidated, and good common storage management processes and procedures are in place for the storage pools identified.
- A storage virtualization initiative budget has been defined.
- There is management comfort with the trade-off between efficiency and loss of vendor choice for the storage pools identified.

Q.2. What do you mean by Data center concepts & requirements?

Ans. A data center (or data centre or datacentre or datacenter) is a facility used to house computer systems and associated components, such as telecommunications and storage systems. It generally includes redundant or backup power supplies, redundant data communications connections, environmental controls (e.g., air conditioning, fire suppression) and security devices.

Data centers have their roots in the huge computer rooms of the early ages of the computing industry. Early computer systems were complex to operate and maintain, and required a special environment in which to operate. Many cables were necessary to connect all the components, and methods to accommodate and organize these were devised, such as standard racks to mount equipment, elevated floors, and cable trays (installed overhead or under the elevated floor). Also, a single mainframe required a great deal of power, and had to be cooled to avoid overheating. Security was important - computers were expensive, and were often used for military purposes. Basic design guidelines for controlling access to the computer room were therefore devised.

Q.3. Briefly Explain Data center classification?

Ans. The simplest is a Tier 1 data center, which is basically a server room, following basic guidelines for the installation of computer systems. The most stringent level is a Tier 4 data center, which is designed to host mission critical computer systems, with fully redundant subsystems and compartmentalized security zones controlled by biometric access controls methods. Another consideration is the placement of the data center in a subterranean context, for data security as well as environmental considerations such as cooling requirements.

The German Datacenter star audit programme uses an auditing process to certify 5 levels of "gratification" that affect Data Center criticality.

Q.4. What are Design considerations for data centers?

Ans. A data center can occupy one room of a building, one or more floors, or an entire building. Most of the equipment is often in the form of servers mounted in 19 inch rack cabinets, which are usually placed in single rows forming corridors (so-called aisles) between them. This allows people access to the front and rear of each cabinet. Servers differ greatly in size from 1U servers to large freestanding storage silos which occupy many tiles on the floor. Some equipment such as mainframe computers and storage devices are often as big as the racks themselves, and are placed alongside them. Very large data centers may use shipping containers packed with 1,000 or more servers each;[8] when repairs or upgrades are needed, whole containers are replaced (rather than repairing individual servers).

Q.5. What is Fire protection?

Ans. Data centers feature fire protection systems, including passive and active design elements, as well as implementation of fire prevention programs in operations. Smoke detectors are usually installed to provide early warning of a developing fire by detecting particles generated by smoldering components prior to the development of flame. This allows investigation, interruption of power, and manual fire suppression using hand held fire extinguishers before the fire grows to a large size. A fire sprinkler system is often provided to control a full scale fire if it develops. Fire sprinklers require 18 in (46 cm) of clearance (free of cable trays, etc.) below the sprinklers. Clean agent fire suppression gaseous systems are sometimes installed to suppress a fire earlier than the fire sprinkler system. Passive fire protection elements include the installation of fire walls around the data center, so a fire can be restricted to a portion of the facility for a limited time in the event of the failure of the active fire protection systems, or if they are not installed. For critical facilities these firewalls are often insufficient to protect heat-sensitive electronic equipment, however, because conventional firewall construction is only rated for flame penetration time, not heat penetration. There are also deficiencies in the protection of vulnerable entry points into the server room, such as cable penetrations, coolant line penetrations and air ducts.

Q.6. How can you calculate Energy efficiency?

Ans. The most commonly used metric to determine the energy efficiency of a data center is power usage effectiveness, or PUE. This simple ratio is the total power entering the data center divided by the power used by the IT equipment.

Power used by support equipment, often referred to as overhead load, mainly consists of cooling systems, power delivery, and other facility infrastructure like lighting. The average data center in the US has a PUE of 2.0, meaning that the facility uses one Watt of overhead power for every Watt delivered to IT equipment. State-of-the-art data center energy efficiency is estimated to be roughly 1.2. Some large data center operators like Microsoft and Yahoo! have published projections of PUE for facilities in development; Google.

Q.7. What is Network infrastructure?

Ans. Communications in data centers today are most often based on networks running the IP protocol suite. Data centers contain a set of routers and switches that transport traffic between the servers and to the outside world. Redundancy of the Internet connection is often provided by using two or more upstream service providers (see Multihoming).

Some of the servers at the data center are used for running the basic Internet and intranet services needed by internal users in the organization, e.g., e-mail servers, proxy servers, and DNS servers.

Network security elements are also usually deployed: firewalls, VPN gateways, intrusion detection systems, etc. Also common are monitoring systems for the network and some of the applications. Additional off site monitoring systems are also typical, in case of a failure of communications inside the data center.

Q.8. What is Disaster recovery?

Ans. Disaster recovery is the process, policies and procedures related to preparing for recovery or continuation of technology infrastructure critical to an organization after a natural or human-induced disaster. Disaster recovery is a subset of business continuity. While business continuity involves planning for keeping all aspects of a business functioning in the midst of disruptive events, disaster recovery focuses on the IT or technology systems that support business functions.

Disaster recovery as a concept developed in the mid to late 1970s as computer center managers began to recognize the dependence of their organizations on their computer systems.

At that time most systems were batch-oriented mainframes which in many cases could be down for a number of days before significant damage would be done to the organization.

Q.9. Basic principles of Security.

Ans. (*a*) Confidentiality: Confidentiality is the term used to prevent the disclosure of information to unauthorized individuals or systems. For example, a credit card transaction on the Internet requires the credit card number to be transmitted from the buyer to the merchant and from the merchant to a transaction processing network. The system attempts to enforce confidentiality by encrypting the card number during transmission, by limiting the places where it might appear (in databases, log files, backups, printed receipts, and so on), and by restricting access to the places where it is stored. If an unauthorized party obtains the card number in any way, a breach of confidentiality has occurred.

Breaches of confidentiality take many forms. Permitting someone to look over your shoulder at your computer screen while you have confidential data displayed on it could be a breach of confidentiality. If a laptop computer containing sensitive information about a company's employees is stolen or sold, it could result in a breach of confidentiality. Giving out confidential information over the telephone is a breach of confidentiality if the caller is not authorized to have the information.

Confidentiality is necessary (but not sufficient) for maintaining the privacy of the people whose personal information a system holds.[citation needed].

(*b*) Integrity: In information security, integrity means that data cannot be modified undetectably.[citation needed] This is not the same thing as referential integrity in databases, although it can be viewed as a special case of consistency as understood in the classic ACID model of transaction processing. Integrity is violated when a message is actively modified in transit. Information security systems typically provide message integrity in addition to data confidentiality.

(*c*) Availability: For any information system to serve its purpose, the information must be available when it is needed. This means that the computing systems used to store and process the information, the security controls used to protect it, and the communication channels used to access it must be functioning correctly. High availability systems aim to remain available at all times, preventing service disruptions due to power outages, hardware failures, and system upgrades. Ensuring availability also involves preventing denial-of-service attacks.

(*d*) Authenticity: In computing, e-Business and information security it is necessary to ensure that the data, transactions, communications or documents (electronic or physical) are genuine. It is also important for authenticity to validate that both parties involved are who they claim they are.

(*e*) Non-repudiation: In law, non-repudiation implies one's intention to fulfill their obligations to a contract. It also implies that one party of a transaction cannot deny having received a transaction nor can the other party deny having sent a transaction.

Electronic commerce uses technology such as digital signatures and public key encryption to establish authenticity and non-repudiation.

OBJECTIVE TYPE QUESTIONS

1. What addressing mode is used in Windows and Sun with Fibre Channel?
   (a) LUN Addressing
   (b) Logical Addressing
   (c) Volume Set Addressing
   (d) Peripheral Device Addressing
2. Which tools are used in the management of the Connectrix Switch?
   (a) Serial Connect, SSH, Embedded Web Server, Web Tools and Connectrix Manager
   (b) Telnet, Embedded Web Server, Web Tools Serial Connect and Connectrix Manager.
   (c) Embedded Web Server, Parallel Connect, SSH, Web Tools and Connectrix Manager
   (d) Parallel Connect, Telnet, Embedded Web Server, Web Tools and Connectrix Manager.

3. How many ports are there in a M-Series ASIC?
   (b) 2 (b) 3
   (c) 4 (d) 8
4. Which command will show the version of PowerPath being run?
   (a) powermt version (b) powermt show version
   (c) powermt display version (d) powermt check_registration
5. Click the Exhibit button.
   What do the lights on the WWN bezel of the Connectrix 12000B indicate?
   (a) power LED / status LED (b) activity LED / power LED
   (c) wwn LED / cabinet temperature (d) fan speed indicator / temp indicator

ANSWERS

1. (d) 2. (b) 3. (d) 4. (a) 5. (s)

❑❑❑

CHAPTER

6

STANDARD APPLICATIONS FRAMEWORK FOR EPICS

6.0. INTRODUCTION

The Standard Applications Framework for EPICS (SAFE) provides an organized approach for building EPICS applications on a Unix host, include the cross-compiled VxWorks parts, the native Unix parts, and ancillary tools. SAFE provides disk directory layouts, scripts, recipes, and documents. A particular feature of SAFE is the separation of the VxWorks/booting and IOC/initialization processes into two phases, each of which takes place in its own *area*, called *IOC* and *Application*, respectively. Each area has a creation script that builds and populates its directories, based on the SAFE configuration file and those from EPICS.

This document describes the procedures necessary for any project to use SAFE to develop and operate IOC applications. Although SAFE proper resides in the Application and IOC areas, its correct functioning is dependent upon some conventions on disk layout, on user/group naming, on network configuration, on installing VxWorks, and on installing EPICS.

Although initially much installation activity takes place in the files associated with VxWorks and EPICS, except for updates these files remain quiescent. The Application and IOC areas refer to them with symbolic links. Most other development activity takes place in the IOC and Application

Inside this Chapter

areas, and all operational activity (actually "driving" the system) takes place in the Application area.

Although this document will cover some aspects of the VxWorks and EPICS installations, such information is intended only to support understanding of creating and using the Application and IOC areas; full installation of VxWorks and EPICS is assumed to be provided by an experienced person.

This version refers to EPICS release 3.13 beta12 or 3.13.0. It will support VxWorks version 5.3.1 for the 680 × 0 and PPC families.

6.1. DIRECTORY LAYOUT

A minimum of 500MB on a dedicated partition is recommended. Unless otherwise noted, the same group is used for all directories (such as "epics") and group-write permission is enabled, so root access is not needed for shared maintenance. Consider then such a partition:

```
/project/  "$PROJECT"
master/
devel/
repository/  "$CVSROOT"
```

where the three main subdirectories are for master copies, development, and the CVS repository. It is convenient to define:

```
setenv  $PROJECT  /project
setenv  $CVSROOT  /project/repository
```

The master subdirectory holds any external sources, such as EPICS, VxWorks, Tcl/Tk, Interviews, and so forth. These are typically not locally configured, but represent read-only originals. The convention is to install each under its generic name, with a subdirectory for the version, for example:

```
/project/master/
    vxworks/
        vxw_5.2/
        vxw_5.3.1/
        .../
    epics/
    tcltk/
    iv/
    .../
```

The development area has the same generic- and version-level structure, except a symbolic link release points to the current release of each package. It also contains the "boot" area which is described fully in the IOC Area and Application Area sections.

```
/project/devel/
    vxworks/
    epics/
        R3.12.2-LBL.6/
        R3.13.0-LBL.1/
        ...
        release@
    boot/
```

Two methods are used to flesh out the development tree: *shadowing* and CVS import/checkout. I use shadowing for VxWorks, since very little is modified: In this method, a symbolic link is made from the version-level directory in devel to its counterpart in master; then the realize script is repeatedly invoked to "drill down" to the few files actually needed for local configuration (RCS can be utilized on a per-file basis). I use CVS for EPICS, since more files are modified. Each method is further described in later sections.

6.2. UNIX ADMINISTRATIVE ISSUES

All IOCs will boot with a unique identity. Initially they use rsh to load the kernel, then NFS for making file accesses after proper mounts are performed. (I do not recommend booting with ftp nor making file accesses with rsh.) In addition to each developer and operator having an account, a special account is created just for the IOCs. In the example following, the names and numbers are, of course, just illustrative; follow normal conventions for your site.

- Create an entry in /etc/passwd:

```
vxuser:*:1000:100:VxWorks:/project/devel/boot:/bin/csh
```

- At least two groups are needed: One for VxWorks installation + maintenance/use; and another for EPICS installation + maintenance/use. (A more elaborate scheme would have four separate groups.)

  Although the IOC account is by default in the VxWorks group, it is also in the EPICS/use group and uses that for NFS activity. Create entries in /etc/group:

```
vxgroup:*:100:maintainer,...
                          (people qualified for VxWorks maintenance)
```

  and

```
epics:*:101:vxuser,maintainer,...
                   (people quali?ed for EPICS maintenance and/or use)
```

- Create the file .rhosts in /home/project/devel/boot with user "vxuser" and group "vxgroup" and permissions -rwxrwxr-w. Put the IP names for all IOCs in it:

```
iocA vxuser
...
```

- The IOC IP address/name pairs should also be entered in /etc/hosts.
- Enable NFS export for /home/project:

```
SunOS, HPUX (/etc/exports);
    /project -access=iocA:...
Solaris (/etc/dfs/dfstab):
    share -F nfs -o rw=iocA:... /project
```

  where iocA:... is the expanded list of IOC names; or a "netgroup" mechanism may be used.

6.3. INSTALLING VXWORKS

Install from the CD-ROM into master/vxworks/<release>. (You will need the installation keys, but not the Tornado license.) Owner should be you or root, group should be "vxgroup", generally every- thing read-only. Then, in devel/vxworks, create the symbolic link ../../mas- ter/vxworks/<release> and using realize, work down the hierarchy ("realize <dir>; cd <dir>"), finally creating only two actual files plus one more for each BSP (where <bsp> = mv177, ..., using standard VxWorks notation):

```
/project/devel/vxworks/<release>/
   target/config/
   all/
      configAll.h
      usrConfig.c
   <bsp>/
      sysLib.c
   <bsp>/
      sysLib.c
   .../
```

At this point I recommend changing the default VxWorks configuration away from the WRS default. It will serve as a template for the local copies made later by iocMakeDir (see IOC Area section) or can even be used directly. For standard EPICS, within configAll.h the following "facilities" should be enabled by moving the #define statements from the "excluded" to the "included" section:

```
INCLUDE_CONFIGURATION_5_2      /* pre-tornado tools */
INCLUDE_NFS                    /* nfs package */
INCLUDE_PING                   /* outbound ping utility */
INCLUDE_RDB                    /* remote debugging (old) */
INCLUDE_SECURITY               /* shell security */
INCLUDE_POSIX_ALL              /* all POSIX functions */
```

You may optionally use the following for bundled C++:

```
INCLUDE_CPLUS                  /* C++ support */
INCLUDE_CPLUS_IOSTREAMS        /* C++ iostreams classes */
```

To enable use of windSh (which includes Tornado-style debugging) yet retain the kernel symbol table, loader, and so forth, as required by EPICS, use:

```
INCLUDE_SYM_TBL_SYNC           /*synchronize host and target symbol tables*/
```

- INCLUDE_CONFIGURATION_5_2 defeats "Tornado" and reverts to the simple approach the EPICS community has been using for 10 years. For now, it is required for EPICS. It means the shell and symbol table are on the target, not on Unix. (It also means you don't need the host-based licensing scheme!) This "facility" also enables rlogin, telnet, some diagnostics, and usage of a start-up script (see EPICS section).
- INCLUDE_NFS is required by SAFE, greatly speeds things up, and enables other features. It depends on the NFS export (see Unix Administration section).
- INCLUDE_SECURITY is one I insist on, especially with rlogin and telnet enabled. Although the default username/password are well known to WRS customers they are not likely known to the general Internet community. I recommend that each project choose its own standard; these is done by editing the #define's in configAll.h (vxencrypt must be run in a shell):

```
LOGIN_USER_NAME  "target"
LOGIN_PASSWORD  "bReb99RRed"   /* "password" after vxencrypt*/
```

- INCLUDE_POSIX_ALL is also required by EPICS and encourages portable code.
- INCLUDE_SYM_TBL_SYNC is available only in version 5.3.1 or later. It keeps the in-kernel and windSh symbol tables synchronized. (You must separately license the windSh, configure it properly, and initiate the target server with "-s" option.) With this option, the older RDB is automatically replaced by the

newer WDB. For VxWorks 5.3 or earlier, which lacked this feature, windSh and EPICS are incompatible.

- usrConfig.c is where any processor registers need to be changed, for example, setting up Transparent Translation registers, special DMA, etc.
- sysLib.c is where the default mapping from VME space to processor space is kept. WRS by default maps A16, A24, and a limited amount of A32 (see Appendix C for details).

It is worth building a kernel and booting it directly, even though the full SAFE recapitulates the process locally (see IOC Area section). You must "source" the following, which you can simplify for a single OS1.

You may quickly synthesize it from site/vxwSetup and start-up/Site.cshrc in the EPICS area if you have already performed part of that installation (see EPICS section)2. setenv WIND_BASE $PROJECT/devel/vxworks/release if ("$ARCH" == "solaris") then setenv WIND_HOST_TYPE sun

```
  4-solaris2
else if ( "$ARCH" == "sun4" ) then
  setenv WIND_HOST_TYPE sun4-sunos4
else if ( "$ARCH" == "hp700" ) then
  setenv WIND_HOST_TYPE parisc-hpux9
else
  setenv WIND_HOST_TYPE unknown
  echo Warning: WIND_HOST_TYPE not set!
endif
setenv PATH ${WIND_BASE}/host/${WIND_HOST_TYPE}/bin:${PATH}
setenv MANPATH ${WIND_BASE}/host/man:${MANPATH}
setenv MANPATH ${WIND_BASE}/host/${WIND_HOST_TYPE}/man:${MANPATH}
setenv MANPATH ${WIND_BASE}/target/man:${MANPATH
```

If the path is right, "which make" will find the one in host/$WIND_HOST_TYPE/ bin/make. "make" should build a few .o's, some other files, and then VxWorks and VxWorks.sym. These latter two ?les are what is actually "booted."

You may wish to verify the VxWorks installation now, or wait until you have created an IOC area. The advantage of the latter choice is that the non-volatile RAM (NVRAM) parameters are derived for you. In any case, you should install the VxWorks boot ROMs or program the flash-RAM as appropriate for your board type.

In order to set up the NVRAM, you must connect to the console port of the VME board, either from a terminal, from a terminal server, or from the serial port of a workstation using the tip facility, kermit, or equivalent.

At power-on or reset you should see:

```
Press any key to stop auto-boot...
[VxWorks Boot]:
```

Try "help" and familiarize yourself with the print ("p"), change ("c"), and boot ("@") syntax. Then change the parameters to something like:

```
boot device         : ei
processor number    : 0
host name           : myhost
```

1. I always set $ARCH early in the login process.
2. The example is correct for VxWorks 5.3.1. Earlier versions, such as 5.2, require a different setup.

3. Sun systems typically have a reference to the "B" serial port called "hardwire."

```
   file name                : /project/devel/vxWorks/release/target/config/<bsp>/
vxWorks
   inet on ethernet (e)     : 127.0.0.1:ffffff00
   inet on backplane (b)    :
   host inet (h)            : 127.0.0.2
   gateway inet (g)         :
   user (u)                 : vxuser
   ftp password (pw) (blank = use rsh):
   flags (f)                : 0 x 0
   target name (tn)         : my Target
   startup script (s)       :
   other (o)                :
```

where your "host name", "host inet", and "netmask" can be found on a Sun, for example, with: "ifconfig le0"; and your target's "inet on ethernet" can be found from "nslookup myTarget". The tricky bit is that the "inet-on-ethernet" really requires the inet and netmask to appear on the one line with the ":" between them. The "user" should correspond to the owner of the .rhosts file (see Unix Administration section). Also, inspect the output of "ifconfig le0" to ensure that BROADCAST is enabled (for EPICS).

You should get a successful boot with a banner and then the shell prompt: ">". The date on the banner should match the vxWorks you built. rlogin, telnet, ping, and so forth should work. Note, typically, the absence of an entry for "gateway-inet." If the Unix host you are booting from is not on the same "wire." that is, is not connected solely by hubs and switches, you will have to enter the router's IP address there. If the "vxuser" is legitimate and you have a properly prepared the ?le ".rhosts" vxWorks will use rsh for accessing the host ?le system it booted from (ls, cd, and so forth work).

6.4. INSTALLING EPICS

The naming scheme is "R<APS-release>-<SITE>.<version>", for example: "R3.13.0-LBL.1". (The LBL EPICS is slightly modi?ed from a standard APS release; this document describes only the LBL type.)

The upper directory layout is:

```
/project/master/epics/
<version>/
    /
```

and

```
/project/devel/epics/
<version>/
```

1. As a general security precaution, note that as I have set things up, access to the VxWorks prompt gives much read and some write access to your Unix file system. I strongly recommend not ever adding a default route, thus defeating most access from hosts not on your "wire." If you are in a site where many offices and building share one large, bridged network, you should consider creating a small, isolated subnet with just the IOCs and selected hosts on it.

In any case, if you need to allow TCP access from beyond your nearest router, use a hostAdd and routeAdd just for specific hosts. Templates for this are provided by the installation script for the IOC area.

```
../
    release@
```

where release points to the current release. Within each version are five main subdirectories (site is found only in the LBL types and in particular supports SAFE itself):

```
startup/
config/
base/
extensions/
site/
```

The installation is based on CVS. I recommend at least CVS version 1.6 and prefer 1.9. To aid in maintaining an audit trail, I add to $CVSROOT/CVSROOT/loginfo:

```
ALL /usr/ucb/mail -s %s maintainer,...
```

To aid in import, checkout, and export it is convenient to add to $CVSROOT/CVSROOT/modules:

```
base        epics/base
config      epics/config
extensions  epics/extensions
site        epics/site
startup     epics/startup
devel       - a base config site startup extensions
```

- Normal installation typically begins with extraction of a "tar"-format archive within the master version directory. Then, using the LBL designation for "<version>", for example, "R3.13.0-LBL.1" everything is imported into CVS:

  ```
  d /project/master/epics
  mkdir R3.13.0-LBL.1; cd R3.13.0-LBL.1
  tar xf <tar-file>
  cvs import epics R3-13 R3-13-0-LBL-1
  ```

- The next step is to checkout a working copy, using a local designation for "<version>", for example, "R3.13.0-SITE.1":

  ```
  cd /project/devel/epics
  cvs checkout -d R3.13.0-SITE.1 -N epics
  ```

  which creates the directory as well. Since the "extensions" to EPICS are not directly tied to a release or version, which really describes the "base," sometime it is useful to only checkout a portion of EPICS: cvs checkout -d R3.13.0-SITE.1 -N startup site base
- At this point you must turn aside from EPICS itself and install the following auxiliary programs:
  - Sun C/C++ unbundled compiler: /usr/lang on SunOS and /opt/SUNWspro on Solaris. If you cannot obtain the Sun compiler, it is possible to copy binary versions of all of the EPICS Unix executables from LBL to their installation directories, however, in some cases (such as Tcl/Tk), they will only run when placed in exactly the same path). Gnu C/C++ (gcc/g++) will build all of base and most of extensions. /usr/ucb/cc does not work;
  - X11R6 and Motif (bundled with Solaris); X11R5 is acceptable;
  - **Tcl/Tk 7.4/4.0:** I maintain sources for Tcl-7.4, Tk-4.0, Blt-1.8, and Tcl-dp3.3b1. No other combination is known to work. See Appendix A for installation recipe;

- **Interviews:** special "LBL for EPICS" version with headers and pre-built lib*.a's. See Appendix B for installation recipe;

- Now 10 files must be tailored for your site (and four more for each additional host type); you will need to have available much of the information you have used in the previous VxWorks and EPICS installation steps. Each tailorable file is named ro.<file>; it should be copied to <file> for modification. ("ro" means "read-only; this keeps the CVS changes isolated from required site dependencies.) In particular, tailor:
  - startup/Site.cshrc and Site.profile; de?ne host ("HOST_ARCH") for Unix building; host-type ("WIND_HOST_TYPE") for VxWorks building;
  - **config/CONFIG_SITE:** defines location of base and extensions;
  - **config/CONFIG_SITE.Unix.<host>:** defines X, Tcl/Tk, Motif, Interviews paths;
  - base/config/CONFIG_SITE: de?nes VME board types; compilers; location of base and extensions;
  - **base/config/CONFIG_SITE_ENV:** defines timezone; IP addresses for NFS, NNTP, LOG servers;
  - **base/config/CONFIG_SITE.Host.<host>:** defines gnu, Tcl/Tk paths;
  - **base/config/CONFIG_SITE_HOST_ARCH.<host>:** defines Perl path;
  - **site/vxwSetup:** defines VxWorks paths;
  - **site/CONFIG_SITE_SAFE:** configures SAFE (follow the included comments closely and see Configuration section);

> **Note:** it is neither necessary nor desirable to have multiple copies of "extensions" when there are multiple copies of base; CONFIG_SITE in each release does the coupling. (Of course, a symbolic link will work as well.)

> **Note:** "base" will contain executables for all hosts that share the file system (for example, SunOS and Solaris-"sun4" and "solaris") and loadables for all targets for which you have BSPs. Do not checkout separate versions for this reason!

- "source startup/Site.cshrc; source site/vxwSetup". This step should be done **once** for each shell before performing any other actions with regard to EPICS in that shell.
- "cd base; "gmake". It will build the host executables and target loadables. (For later auditing, I recommend redirecting output: "gmake >&makelog".) Repeat on alternate host types if appropriate; use "gmake install.host" since the target loadables were built in the first step. If you have copied the Unix executables because you do not have the unbundled Unix compiler, tailor config/ CONFIG_BASE as directed by the comments within it; alternatively, use "gmake install.cross".

> **Note:** The Unix executables in "base" are con?guration tools, **not** Channel Access clients.

I have tailored the "base" of EPICS to build a large and useful set of record, device, and driver object files; further customizing and selection for each application is handled by procedures described in the Application Area section. Note: Certain assumptions about where various VME modules appear in the VME address spaces are built-in (see Hardware Setup section).

- "cd ../config; ./bldCfgLnks"; makes symbolic links to "base/config" so site-dependent items do not need to be duplicated for "extensions".
- "cd ../extensions; gmake >&makelog". (Omit if you have copied the Unix executables because you do not have the unbundled Unix compiler; in fact, do not checkout the source from CVS.) As does "base", "extensions" will contain executables for all hosts that share the file system, so repeat the procedure on all host types.

> **Note:** Only a subset of extensions are called out; with the exception of "BURT," these are just the ones listed in my Web page "Recommended EPICS documents." You can tailor the subset in config/CONFIG_EXTENSIONS.

You may notice that vxWorks is copied into the loadables area. It is not used by SAFE.

- "cd ../site; ./bldEnvData"; extracts much detail from "base" and "config" and creates "resource.def" and "envSet" which will be used in the Application Area section. This is strictly for SAFE.
- "cd ../base/bin/<host>;
  ln -s ../../../extensions/bin/<host>/gdct313 gdct;
  ln -s ../../../extensions/bin/<host>/dct313 dct"; places simpler references to these con?guration tools (which are not Channel Access clients) into "base" with like tools.
- Host startup. There are two host executables intended for operating EPICS:
  - iocLogServer. There should be one of these running as a Unix daemon for the whole site. Typically an initialization line is added to rc.local or equivalent for a selected workstation (details upon request). It provides a very primitive global occurrence logger. By default, this capability is disabled in the IOCs by SAFE.
  - caRepeater. There should be one of these running as a Unix daemon for each workstation. Usually, the first instance of any Unix Channel-access client will cause caRepeater to be spawned (details upon request).

6.5. IOC AREA

6.5.1. Facilities of IOC Area

The IOC area provides the following facilities:

- Linking to particular versions of EPICS and VxWorks is mostly driven from EPICS configuration files by means of a special script "iocMakeDir".
- The boot lines stored in the NVRAM of the VME board never change. An initial set is derived for each IOC based on configuration files.
- Each "logical" IOC-a combination of a particular board type plus its usage-has its own area.
- Switching between applications or combining several applications on one IOC is easily done.
- The VxWorks kernel can be customized for each IOC without altering the master installation, yet site-wide configurations may be shared.
- A per-IOC writable directory is easily accessible from applications which can thus use generic file names.

- Several IOCs can run same the same application relatively easily.
- Tailoring of network parameters that are common to all IOCs is kept in one place An initial set of commands is derived based on configuration files.

6.5.2. Detailed Layout

```
/project/devel/boot/iocs/
    all/
        netInit.cmd
        "/vxboot"
    iocA/
        IOC_iocA
    log/
    cmd/
        example.cmd
        none.cm
        ...
    appinit.cmd@
    configVx/
        all/
            configAll.h
            usrConfig.c
            ...
        <bsp>/
            sysLib.c
            ...
            vxWorks
            vxWorks.sym
        .../
    vxWorks@
    vxWorks.sym@
    NVRAM.<bsp>
    startup.cmd
    "/log"
iocB/
.../
[links to EPICS and VxWorks]
iocB/
.../
```

The all area is shared among IOCs. netInit.cmd has host additions, routing additions, and NFS mounts which refer to the host or to common network parameters. The primary mounts are:

- /vxboot to refer to IOC area root;
- /log to provide generic (writable) logging area per-IOC;
- /apps to refer to Application area root (see Application Area section).

6.5.3. Usage

An IOC area is built by invoking from within the iocs root directory the script:

```
<path-to>/iocMakeDir [-v] [-LINK] <ioc> <bsp>
```

where,

- -v optionally requests verbose output;
- -LINK optionally makes symbolic links for vxWorks directly to the VxWorks installation rather than building a copy of the kernel locally; by default, iocMakeDir is in the COPY mode;
- <ioc> is the name of the IOC: iocA, iocB, ... in the layout example;
- <bsp> is one of the VxWorks standard BSP designations (mv177, ...).

  iocMakeDir is configured by the EPICS configuration files config/CONFIG* and additionally by site/CONFIG_SITE_SAFE (see Configuration section). It is called with an absolute path so that it may "lock" to a specific version of EPICS. iocMakeDir creates and/or fills in directories and files. Note carefully all informational, warning and error messages and take appropriate action. iocMakeDir never over-writes actual files, thus, it may be re-run over the same area to replace missing files or, by first removing them, to update files. (The symbolic links to EPICS are replaced-a way of switching EPICS versions. If this is your purpose, any unmodified files should be deleted, so you do get the latest versions, and any modified ones renamed for subsequent merging.)
- The first required argument is an existing or new IOC name. If the name already exists, then missing files or subdirectories are filled in; no actual files are over-written. A typical usage would be to change the BSP (target) type. (Changing the BSP will leave previous versions intact.)
- The second required argument specifies the BSP (target) type. The latter must refer to a target which you have called out in the CROSS_COMPILER_TARGET_ARCHS line in config/CONFIG_SITE and which has undergone the EPICS build process.
- If you wish to have all IOCs use identical kernels as built in the VxWorks area, specify -LINK as an optional argument. LINK defeats the ability to have variant kernel configurations for each IOC and requires that the VxWorks distribution itself be NFS mounted by netInit.cmd (as described below). With LINK, there is no local copy of vxWorks, just a symbolic link back to the kernel you built in the VxWorks area (devel/vxworks/target/config/<bsp>). Be sure that kernel is properly configured and built.
- If you want a verbose commentary as iocMakeDir runs, specify -v as an optional argument.

  iocMakeDir builds the directory structure such as "iocA" above, and populates it with links to the EPICS area. It creates the subdirectories cmd and configVx and the files startup.cmd, IOC_<ioc> and NVRAM.<bsp>.
- IOC_<ioc> is a 'marker' used by iocMakeDir.
- NVRAM.<bsp> is a prototype of the entries to be made in the non-volatile ram (NVRAM) area of the IOC, either at the boot ROM prompt (see VxWorks section) or from a running VxWorks system, using bootChange. The accuracy of it is dependent on the correctness of site/CONFIG_SITE_SAFE.

  If you previously set the NVRAM up for booting directly from the VxWorks installation, you should now change the two lines "file name" and "startup script" after carefully checking the prototype.
- startup.cmd is the initial set of shell commands for the IOC. Again, carefully check it.

If the -LINK option was not used when iocMakeDir was invoked, iocMakeDir populated the configVx subdirectory with copies of the relevant files from the VxWorks area (configAll.h, usr-Config.c, sysLib.c, and others); these can be tailored for just this IOC1. To build the kernel (even with no tailoring):

```
source startup/Site.cshrc
source site/vxwSetup
cd configVx/<bsp>
make
```

A number of .o files and others will be created, then finally vxWorks and vxWorks.sym.

Changing the BSP will leave configVx/<bsp> directories intact. You can revert quickly to a previously built BSP by either re-invoking iocMakeDir, or by switching the links vxWorks* (and possibly appinit.cmd).

1. There are links to other files which, in general, should not be modified. Should you decide to do so, realize each file first.

The first invocation of iocMakeDir within the IOC area will also build a shared subdirectory all.

The netInit.cmd there should be carefully checked; its accuracy is dependent on the correctness of site/CONFIG_SITE_SAFE.

- Be sure the paths for NFS mounts "/apps" and "/vxboot" are correct-subsequent scripts depend on them.
- The templates for two other NFS mounts are left disabled. However:
  - The "epics" NFS mount may need to be enabled if the default (LINK) option was taken with app-Makedir, in order that the link target<bsp> ~> base/bin/<bsp> ~> <path-to-epics- release>/base can be resolved from the perspective of VxWorks (see COPY option in Application section).
  - The "vxworks" NFS mount may need to be enabled if the LINK option was used in iocMakeDir, in order that the link vxWorks ~> <path-to-vxworks-version>/target/config/<bsp>/vxWorks can be resolved from the perspective of VxWorks.
- Templates are provided for adding hosts and routes for machines not on the local "wire."1(See the last part of the VxWorks Installation section for more details.)

appinit.cmd is a link, used by startup.cmd, to proceed automatically from the VxWorks primary boot process directly into loading an EPICS application. The cmd subdirectory contains small auxiliary command files to act as "bridges." Initially, appinit.cmd is linked to none.cmd, a null file, which ends the boot process cleanly with no EPICS application running. Also provided is a template, example.cmd, which should be copied and tailored for any actual application(s). Then, to move between no application or any particular application requires changing only the link appinit.cmd; neither the NVRAM nor the master startup.cmd file needs to be edited. From the null state, an application can easily be loaded manually; at the VxWorks prompt:

```
> cd "/apps/myApp"
> <cmd/startup.<bsp>.cmd
```

The uid/gid in the nfsAuthUnixSet line should match the "vxuser" passwd and group entries. When combined with the Unix permissions, the IOC should be able to read all the mounted areas and write into /log; make sure group-write is set on the latter and that the group is correct.

6.6. APPLICATION AREA

The Application area provides the following facilities:

- Linkage to EPICS and VxWorks driven from EPICS configuration by means of special script app-MakeDir.
- Builds for multiple targets.
- Incremental building supported.
- Leverages heavily from EPICS Makefiles-no rules needed, just lists
- Customized record, device, and driver support provided, leaving main EPICS unaltered:
  - allows minimal loading on per-application basis;
  - easy to try new/modified types.

1. With later versions of VxWorks, if the IOC is booted through a gateway (router), the appropriate entries are made automatically, hostShow and routeShow are useful in confirming this.

Most of the "driving" activity takes place in the Application areas: configuring records, screens, alarms, archiving, etc, and then actively manipulating hardware from a running IOC.

6.6.1. Detailed Layout

```
/project/devel/boot/apps
  "/apps"
  appX/
     APP_appX
     Makefile
     db/
     src/
     rec/
     dev/
     drv/
     dbd/
     cmd/
     resource.def
     envSet
     dl/
     alh/           [optional]
     ar/            [optional]
     target<bsp>@
     ...
     [links to EPICS and VxWorks]
  appY/
  .../
```

6.6.2. Usage: Create phase

An Application area is created by invoking from within the apps root directory the script:

```
<path-to>/appMakeDir  [-v]  [-COPY]  <app>  [<target> ...    |ALL|NONE]
```

where,

- -v optionally requests verbose output;
- -COPY optionally copies generic EPICS files rather than linking to them; by default, appMakeDir is in the LINK mode;
- <app> is the name of the Application area (appX, appY, ...in the layout above);
- <target> is an optional list of one or more VxWorks standard BSPs (mv177, ...); or ALL; or NONE.

  appMakeDir is configured by the EPICS configuration files config/CONFIG*, base/con-fig/CONFIG*, and additionally by site/CONFIG_SITE_SAFE (see Configuration section). It is called with an absolute path so that it may "lock" on to a specific version of EPICS. appMakedir creates and/or fills in directories and files. Note carefully all informational, warning and error messages and take appropriate action.
- The COPY option replaces links with copies of the generic EPICS files iocCore, seq, init-HookLib, and devLibOpt. Each symbolic link target<bsp> becomes a subdirectory into which the files are placed. This option is useful if the EPICS master area will not be accessible to VxWorks by an NFS mount using that symbolic link during EPICS loading (for example, if EPICS is in an AFS or DFS area). Note carefully that the default LINK option, may require manually configuring an NFS mount in the all/netInit.cmd file in the IOC area (see IOC area section).
- The first required argument is an existing or new Application name. If the name already exists, then missing files or subdirectories are filled in; no actual files are over-written. A typical usage would be to add another BSP. However, the symbolic links to EPICS are replaced-a way of switching EPICS versions. (If this is your purpose, any unmodified files should be deleted, and any modified ones renamed for subsequent merging; and typically then you would use the "NONE" argument for the target type.)
- The second required argument specifies one or more BSP (target) types. Each of these must refer to a target which you have called out in EPICS in the CROSS_COMPILER_TARGET_ARCHS line in config/CONFIG_SITE. If "ALL" is specified, the entire list in this line is used. If the argument is NONE the BSP lists and files are left intact-this is useful when upgrading an existing Application area to a new version of EPICS, or when replacing missing or damaged files with originals.

The fundamental application is developed in the db and src subdirectories:

- db is where the configuration of the "function-block" database is prepared. This is the list of "records" and their non-default parameters that comprise the heart of an EPICS application. A variety of techniques are available. If gdct ("Graphical Database Configuration Tool) or dct are used, they will key off the defining <app>.dbd file; a link is pointed to the actual file (see Build instructions, below). As explained in the Configuration section, provision is also made for supporting CAPFAST.
- When needed, "C" code for the functional parts of subroutine records, and SNL code for sequencers, is prepared in the src subdirectory. A template Makefile is provided; follow the comments to specify your actual source and object files; the built-in EPICS rules are sufficient for typical usage. Any stand-alone VxWorks tasks may easily be accommodated here, as well as any Unix programs.

The rec, dev, drv and dbd subdirectories are populated with files copied and slightly modified from within the master EPICS area. These four together perform two functions: "tailoring" each Application area to use just those existing records and devices/drivers needed; and "customizing" each Application by modifying or adding new records and devices/drivers, without affecting the master EPICS.

Initially, rec, dev, and drv contain only recBaseLIBOBJS, devBaseLIBOBJS, and drv BaseLIBOBJS, respectively. These are modified from the EPICS originals by disabling all "hard-ware" types, leaving only "software" types. The loading of EPICS requires, ultimately, three additional components in addition to the four in the target<bsp> group: recLib, devLib, and drvLib.

These are built-up by pre-linking the individual VxWorks object modules from the master EPICS area locally. To add standard hardware types already installed in EPICS, the appropriate lines in the respec-tive {rec,dev,drv}BaseLIBOBJS are enabled (by removing the comment symbol and whitespace). To add custom types, the modi?ed or new source code is prepared locally (for example, xxx.c), and one of the following three methods used:

1. add to Makefile.Vx the lines
   LIBOBJS += xxx.o ...
   RECTYPES += xxx.h ... [rec only]
2. add to Makefile.Vx the line
   include xxxBaseLIBOBJS
   and create xxxBaseLIBOBJS with the lines
   LIBOBJS += xxx.o ...
   RECTYPES += xxx.h ... [rec only]
3. add directly to {rec, or dev, or drv}BaseLIBOBJS the lines
   LIBOBJS += xxx.o ...
   RECTYPES += xxx.h ... [rec only]

(Internal comments mark the appropriate places.) The modified or new object module will appear only locally. Using the built-in EPICS RULES, gmake will perform the actions (see Build phase, below).

Note that this action must be perfectly synchronized with modification of the appropriate dbd/*.dbd file.

- dbd provides database definition files that are put through a series of special "build" processes, producing a master file called <app>.dbd. This serves two purposes: it guides the initialization of EPICS in VxWorks so just those records, devices, and drivers called out are initialized, using the specified default field values, by dbLoadDatabase(); and it constrains the database construction tools under Unix (gdct and dct) to allow only the desired entities, to prompt with only allowed options, and to produce only the non-default fields used by the dbLoadRecords() and dbLoadTemplate() procedures used to complete the initialization of EPICS.

  There are five files, initially copied and modified, in dbd:
  - **baseRec.dbd:** calls out records to be included in recLib. Note that any changes must be perfectly synchronized to changes in rec/Makefile.Vx and/or rec/recBaseLIBOBJS.

- **baseDev.dbd:** calls out devices to be included in devLib. Note that any changes must be perfectly synchronized to changes in dev/Makefile.Vx and/ or dev/devBaseLIBOBJS.
- **baseDrv.dbd:** calls out drivers to be included in drvLib. Note that any changes must be perfectly synchronized to changes in drv/Makefile.Vx and/ or drv/drvBaseLIBOBJS. (Special note: it is not possible to specify no drivers at all, so drvDummy exists to make things work; if one of more real drivers are called out, it may be disabled.) Where possible, drivers have been bundled with their devices, so many hardware types appear only in baseDev.dbd.
- **baseBpt.dbd:** calls out breakpoint tables to be included in devLib. For customization, follow the prescription in Chapter 2 of "IOC Application Developer's Guide".
- **baseMenu.dbd:** calls out menu enumerations to be included in devLib. For customization, follow the prescription in Chapter 2 of "IOC Application Developer's Guide".

To add standard hardware types already installed in EPICS, the appropriate lines in the respective {rec, or dev, or drv}.dbd are enabled (by removing the comment symbol and any whitespace). To add custom types, the modified or new database definition is prepared locally (for example, xxx.dbd), and one of the following three methods used:

1. add to each of base{rec, or dev, or drv}.dbd the respective lines
   record ... or
   device ... or
   driver ...
2. add to xxxInclude.dbd the lines
   include recxxx.dbd or
   include devxxx.dbd or
   include drvxxx.dbd
   and create xxx.dbd with the respective lines
   record ... or
   device ... or
   driver ...
3. add directly to <app>Include.dbd the lines
   record ... or
   device ... or
   driver ...

(Internal comments mark the appropriate places.) <app>Include.dbd serves as a master organizing file for gmake, allowing (with the Makefile.Host)all of the above possibilities. There is considerable flexibility, and the various inclusions may be mixed in any order that makes maintenance easier.

- cmd subdirectory holds the master startup file called from the IOC area; it calls five auxiliary files in turn.
- resource.def and envSet: these are directly constructed from the configuration files of EPICS, and are used, respectively, for initializing the pseudo-environment variables of EPICS in VxWorks; and for initializing the shell environment when "driving" the Application area (see Build phase, below). odl (and optionally alh, ar) subdirectories: hold the configuration files specific to the edd/dm (and optionally alh, ar) tools. Some initial templates are provided:

- in dl, a link to a suggested Color.adl containing a small, documented set of colors, and a few useful sets of color dynamics rules;
- in ar, a subdirectory arReq containing three sample request files and an initialization file open.dat; plus a logging directory arLog;
- in alh, a sample configuration file.

6.6.3. Usage: Build Phase

Before any further activity, and when subsequently restarting activity, in each such shell, do once: source site/epicsSetup.

This will pick up the local environment variables, and will also conditionally "source" the VxWorks setup file (site/vxwSetup) and the EPICS setup file (startup/Site.cshrc) in the proper order.

It is important to understand that this initialization activity is context-dependent, that is, it must be done from within the specific Application Area you are going to use. It refers to the symbolic links which bind this Application Area to a particular version of EPICS + VxWorks. It sets the customized, synchronized "environments" defined by envSet and resource.def for the Unix-client and VxWorks-server, respectively, that define this application. It can also be done from within the "site" directory of an EPICS release; this would set up for a generic or default Application. The process will not work from within your "home" .cshrc or .login, since 1) the symbolic links are not there; and 2) even if you put in absolute paths or duplicate links you are binding to just one application, EPICS, and VxWorks for all shell windows.

- Site.cshrc sets the HOST_ARCH and WIND_HOST_TYPE environment variables; thus it is needed by both build and run phases of appMakeDir, which will attempt to call it if you have not done so.
- vxwSetup sets the WIND_BASE environment variable and puts the VxWorks tools and manual pages in your path.
- epicsSetup sets the environment variables of envSet; and it puts the base tools (gdct, snc, etc), extensions (edd, dm, alh, ar, etc) and EPICS manual pages in your path. It does this so everything is accessible from within the main Application directory and any first-level subdirectory. This is convenient for invoking gdct in dbl, edd or dm in dl, etc. or any of them from the main directory. epicsSetup calls Site.cshrc and vxwSetup if you have not already done so, thus you may simple use it as a one-step setup.

To actually build or rebuild the tailored or customized EPICS components and any "C" or SNC code you have written, you may either invoke (with no arguments): "gmake" from the top level (recommended once immediately after the Create phase); or individually, gmake to incorporate any changes in rec, drv, dev, dbd, or src from within the respective subdirectories. Note that "gmake" will build for the host type you invoke it from and all target types[1].

- Before loading your application into the IOC, you must modify some of the application-specific files in cmd:
  - epicsCode.<bsp>.cmd should not be modified;
  - userCode.<bsp>.cmd to load your subroutine record procedures, sequencers, and any stand-alone tasks from bin/<bsp>;[2]
  - db.cmd to load and expand your db/*.db and *.tmpl files using dbLoad Records() and dbLoadTemplate() as needed;

- epicsInit.cmd to perform EPICS housekeeping and to select EPICS features such as logging, hot-start, and time-synchronization;
- userInit.cmd to initiate your sequencers and/or to start any stand-alone tasks.

A master coordination file startup.<bsp>.cmd calls the above in order; it should not be modified for a single application. This is the file called by appinit.cmd from the IOC area to proceed from the VxWorks boot to the EPICS initialization. Note that this file is BSP-dependent.

For multiple applications, you must carefully interleave the several instances of db.cmd, user Code.<bsp>.cmd, and userInit.cmd with one invocation of epicsCode.<bsp>.cmd and epicsInit.cmd; building a combining master coordination ?le in a separate area is one method.

6.6.4. Usage: Run Phase

When actually "driving" the Application, you should be in its top directory. Specific OPI tools, such as edd/dm, alh, etc, typically have sufficient path definitions or symbolic links such that they will run from there or from their unique subdirectory.

- dm, in particular, honors the environment variable EPICS_DISPLAY_PATH. This ":" delimited list specifies directories to search for screens. To allow sharing of some screens across Applications, append additional dl directories to it. The default setting allows for finding screens in the local application:

```
.:./dl:../dl
```

1. You may constrain gmake by giving it the argument <goal>.<arch>, where <goal> is either install or build and <arch> is a Unix host or VxWorks BSP (such as solaris or mv177) or a generic class (such as host or cross). Example: gmake build.mv177 or gmake install.host.
2. Every <bsp> type must be so modified.

To find more screens, say, in the Network Statistics Application, change it (in envSet) to:

```
.../dl:../dl:../stats/dl:../../stats/dl
```

- When the IOCs and OPIs ("consoles") are not all on the same subnetwork, then the OPI tools and IOC channel access servers cannot "find" each other. The method of finding is controlled by EPICS_CA_ADDR_LIST. You will have to modify the environment for both categories as suggested below.

To enable an IOC to reach OPIs on other subnets, add their IP addresses to EPICS CA ADDR LIST in resource.def. You must also add a hostAdd() and a routeAdd() to the startup.cmd sequence, typically in cmd/epicsInit.cmd for an isolated IOC or in all/netInit.cmd for all IOCs. (An exception to the latter requirement is that if an IOC is booted with a gateway, the host/route entries are automatically entered for the gateway.) If a router that serves a distant subnet allows it, then you can enable access to all the OPIs on that subnet by using the subnet address in EPICS_CA_ADDR_LIST and just the routeAdd().

Do NOT use the defaultRoute() feature, as it presents a serious security breach. Assume IOC 111.222.33.44 wants to reach OPI 111.222.3.4 and the router on the IOC subnet is 111.222.33.1. Then either:

```
EPICS_CA_ADDR_LIST DBF_STRING 111.222.3.4
routeAdd "111.222.3.4","111.222.33.1"
hostAdd "remoteOPIName","111.222.3.4"
```

or

```
EPICS_CA_ADDR_LIST DBF_STRING 111.22.3.0
routeAdd "111.222.3.4","111.222.33.1"
```

To enable an OPI to reach IOCs on other subnets, add their IP addresses to EPICS_CA_ADDR_LIST in envSet. If a router that serves a distant subnet allows it (a feature called "directed broadcast"), then you can enable access to all the IOCs on that subnet by using the subnet address in EPICS_CA_ADDR_LIST. Assume OPI 111.222.3.4 wants to reach IOC 111.222.33.44. Then either:

```
setenv EPICS_CA_ADDR_LIST 111.222.33.44
```

or

```
setenv EPICS_CA_ADDR_LIST 111.222.33.0
```

6.7. CONFIGURATION

The file CONFIG_SITE_SAFE determines many parameters used by the scripts iocMakeDir and by appMakeDir when building new areas; these scripts also use parameters contained in the standard EPICS config and base/config directories. Some of the parameters are only advisory, for example, those used in constructing the NVRAM file which suggests how to setup the VME board. The parameters are:

6.7.1. Common

- **EPICS_ROOT:** the root of the EPICS version to which the area will be linked. It must be a directory containing at least the subdirectories: startup, site, config; it is required whenever ioc MakeDir and appMakeDir are not called with an absolute path.

6.7.2. For iocMakeDir

- SAFE_BOOT_HOST is the Unix host for booting IOCs; defaults to current host. It will be used for "host name" in NVRAM and for NFS mounts in startup scripts. Needed only if you work from other, NFS-mounted hosts.
- SAFE_BOOT_IP is the IP address for SAFE_BOOT_HOST. If not supplied, then nslookup is used; if that fails, EPICS_TS_NTP_INET or EPICS_IOC_LOG_INET from EPICS "config" will be used. A value of "127.0.0.1" indicates all of the above failed.
- SAFE_BOOT_TGT_IP is the IP for the VME target board (IOC); NOT to be used if you have multiple targets. If not supplied, then nslookup on <target> is used. A value of "127.0.0.1" indicates all of the above failed.
- SAFE_BOOT_NETMASK is the "netmask" for the local LAN of the target. If not supplied, then ifconfig is used (see "LANIF" section below). A value of "ffffff00" is used if all of the above fail. If iocMakedir will be run from hosts with different LAN interfaces, or from a different LAN from where the targets will boot, this parameter *should* be supplied.
- SAFE_BOOT_PATH is the absolute path to the boot area; if not supplied, the directory above the current working directory is used. A warning is issued if the final resolved path used does not contain the subdirectories apps and iocs.
- SAFE_PROJ is the absolute path to the EPICS and VxWorks area; if not supplied "/project" is the default. It is used to construct suggested NFS mounts for the all/netInit.cmd file.

- SAFE_VX_USER, SAFE_VX_USER_UID, SAFE_VX_GROUP, SAFE_VX_USER_GID are the "user" and "uid" in /etc/passwd, and "group" and "gid" in /etc/group as target will appear to host. (See Unix Administrative Issues section for details about this.)

6.7.3. For appMakeDir

- AUX_DIRS is a list of additional directories used for EPICS tools or other purposes, for example, ar and alh; templates may exist to fill them out.
- CAPFAST: "true" if you want subdirectories filled out within db for typical use of this tool; "false" if you are using GDCT or another method.

6.7.4. LANIF and BOOT_DEV setup

- For each VxWorks BSP (target) type, a typical LAN interface boot device abbreviation is listed under "Specific Boot Devices." If your BSP is not shown, consult documentation or try "help" from the VxWorks boot prompt and see "available boot devices" to add to this list.
- For each host type, a typical LAN interface type is listed. If your host type is listed, confirm manually that the LANIF entry works; change it if required, or add an entry if you have a new host type. For the case where you have several hosts with different LANIF types, this method is fails and the netmask should be supplied (see Configuration section).

6.8. BOOTING

Here is the dynamic thread of the booting sequence.

- startup.cmd will be invoked after the kernel is loaded. It will:
  - tailor the console and login prompts;
  - call netInit.cmd for the common commands;
  - mount /log for all applications
  - call appinit.cmd to load the application(s)
  - ends.
- netInit.cmd:
  - performs the NFS mounts and other common commands;
  - returns.
- appinit.cmd
  - changes to the application subdirectory;
  - invokes the local, target-specific startup.<bsp>.cmd;
  - returns.
- startup.<bsp>.cmd:
  - invokes the local, target-specific epicsCode.<bsp>.cmd;
  - invokes the local, target-speci?c userCode.<bsp>.cmd;
  - invokes the local db.cmd;
  - invokes the local epicsInit.cmd;
  - invokes the local userInit.cmd;
  - returns.

In order to load multiple applications into one IOC, the generic application startup's must be broken apart (see Application area section).

6.9. INSTALLING TCL/TK

Please consult http://csg.lbl.gov/csg.html for a current pointer. Obtain the four gzip'ed "tar"-format ?les for tcl, tk, blt and dp. Create a directory tree as: /home/project/master/tcltk/src/ move into it, and gunzip and extract the "tar"-format ?les, resultingin:

```
tcl7.4/
tk4.0/
blt-1.8
tcl-dp3.3b1
```

- Building tcl: Adjust your PATH to ensure that the unbundled C compiler is found before /usr/ucb/bin/cc.

```
cd tcl7.4
./configure --prefix /home/project/master/tcltk
make; make install
```

- Building tk:

```
cd tk4.0
./configure --prefix /home/project/master/tcltk
make; make install
cd ../../lib
ln -s libtk4.0.a libtk.a
```

- Building blt:

```
cd ../src/blt-1.8
./configure --prefix /home/project/master/tcltk
```

Edit Makefile, changing the de?nition of TK_LIB from -ltk4.0 to -ltk; and changing the de?nition of TCL_LIB from -ltcl7.4 to -ltcl.

```
make; make install
```

- Building dp:

```
cd ../src/tcl-dp3.3b1
```

Edit configure.users, changing the de?nition of:

```
DP_ROOT to /home/project/master/tcltk
TCL_INCLUDE to ../tcl7.4
TCL_LIBRARY to ../tcl7.4/library
TCL_LIBDIR to ../tcl7.4
TK_INCLUDE to ../tk4.0
TK_LIBRARY to ../tk4.0/library
TK_LIBDIR to ../tk4.0
```

Prepend to your PATH /home/project/master/tcltk/bin.

```
./configure --with-cc
make; make install
```

Now the directory structure should look like:

```
/home/project/master/tcltk/
    src/
        tcl7.4/
        tk4.0/
        blt-1.8
        tcl-dp3.3b1
```

```
bin
    tclsh
    ...
lib
    libtcl.a
    ...
include
    tcl.h
    ...
man
```

6.10. INSTALLING INTERVIEWS

Please consult http://csg.lbl.gov/csg.html for a current pointer. Obtain the gzip'ed tar-for- mat "include" file and the two "lib.a" files for each flavor of Unix.

Create a directory tree as:

```
/home/project/master/iv/
    lib/
        O.<arch>
        ...
```

where <arch> follows the EPICS convention for host architecture. Move into the top, and gunzip and extract the "tar"-format "include" file. Then move into each lib/O.<arch>, gunzip the appropriate two "lib.a" files, and rename them, resulting in:

```
lib/
    O.<arch>/
    libIV.a
    libUnidraw.a
    ...
include/
    Interviews/
    Unidraw/
    ...
```

6.11. VME HARDWARE AND DRIVER NOTES

One of the most difficult aspects of EPICS is assigning addresses to modules plugged into the VME backplane. The current scheme, which applies to all EPICS version through 3.12, leaves much to be desired. (A great improvement comes with version 3.13.)

6.11.1. Module_types.h

The information is contained in base/include/module_types.h. All current drivers have been assigned a position in an array; the entries describe the start address, number of modules ("slots"), inter-rupt level, and so forth. This is a common module, loosely maintained by the EPICS community. This piece of code requires a great deal of study before attempting modification:-).

By way of introduction, the major scheme is as follows: the are four major sections, for "ai", "ao", "bi", and "bo" types, respectively. Within each section, there are arrays describing the known board types. The position in the array is called out by a #define which will be recognized by the driver/device code. The arrays must be carefully filled out, respecting the position of the entry you are concerned with. The most important arrays are:

num_cards: the number of cards for which "slots" are expected;

num_channels: the number of channels allowed per card;

interruptible: 0/1 denoting whether the device does not/does support interrupts;

bus: the formal EPICS bus enumeration (VME, CAMAC, A-B, etc);

addrs: the address in the VME short-I/O (A16) space;

memaddrs: the address in the VME standard (A24) or extended (A32) space.

Note that the latter two types are mutually exclusive.

For each VME board type, the address switches are set to one of the "slots" that are pre defined for it; the drivers probe the VME address space during initialization. Boards within a type do not need to be in contiguous slots.

The advantage of this method is that no special initialization calls are made during IOC startup: if a record refers to a device, and the driver is configured in, it will find the hardware. Another advantage is that once non-overlapping addresses have been built-in for all known board types, there will never be an address conflict within a specific VME crate. Each site will have its own version, and these definitions may be over-ridden by adding additional code in module_types.c or even with VxWorks command-shell lines.

6.11.2. VME vs. Processor Addresses

The addresses found in module_types.h are true VME addresses, as set in the switches and jumpers of the I/O boards. From within the code running under VxWorks, however, processor addresses are used.

The relationship between these is set by hardware (such as the VMEchip2 ASIC of the MVME-167/177 series) which connects the CPU board's internal bus to the VME bus, and by software, particularly in a structure in sysLib.c called sysPhysMemDesc. For each IOC, you can modify this structure to arrange just the amount of processor address you want assigned to the A32 ("extended"), A24 ("stan-dard") and A15 ("short I/O") spaces of the VME bus.

In general, you should not change the A16 and A24 mappings at all. In a typical Motorola VME board, A24 has the full 16 MBytes mapped from 0xf000'0000 to 0xf0ff'ffff, and A16 has the full 64KBytes mapped from 0xffff'0000 to 0xffff'ffff. Also, a small, 16Mbytes portion of A32 is mapped from 0x0200'0000 to 0x02ff'ffff. You can map much greater amounts than this (up to nearly 2 Gbytes), but doing so will consume large amounts of real RAM and take a long time to initialize.

You may also change the caching. The VME space should always have caching disabled. By default, WRS sets D32 accesses to A24 devices to be changed to two sequential D16 accesses; for some VME boards, this is just the wrong tactic and must be modified.

You should follow the convention that all EPICS drivers do, and never use absolute processor addresses; use the VME addresses and the function sysBusToLocalAddrs() to translate to processor address. The demonstration application "vmeProbe" is an excellent example of this; you are encouraged to use it for actual hardware testing and as a template for coding. (Notice careful use of the function vme-MemProbe() as well.)

In general, I do not recommend attempting to access the VME bus directly from the VxWorks shell; often you will get a false, negative result merely because you have failed to properly translate to processor space or have not put enough "casts" in to generate the proper data width instructions (D8, D16, D32).

6.12. EXAMPLE APPLICATIONS

A "tar"-format file is provided for each; use appMakeDir to create a new Application area, then extract the file (from within the new Application directory); follow the steps in the HOW-TO file; and build using appMakedir again.

6.12.1. VMEbus Probe

This application has a new record type, "vmeProbe." It is a mechanism for probing the VME address space to aid in integrating new VME boards, verifying switch settings, bit patterns, etc.

The capability is provided to read or write from the A16, A24, or A32 address spaces using D8, D16, or D32 transfers, either single-word or multiple-word. Both value and status are returned, thus, there is no side-effect from probing into unimplemented address ranges.

In db, the new record is instantiated, along with a number of auxiliary "soft" records to act is input and output buffers.

In dl, a screen was built for actual use.

6.12.2. Simulator

This application has two layers: the lower layer uses the simulate feature of EPICS records to bypass actual hardware, allowing a simple CALC record to mimic a power-supply with a slow response time. The upper layer uses a sequencer to control the simulated supply. Two independent supplies are instantiated.

The application also demonstrates the hot-start feature of the IOC: All dynamic parameters are restored after a reboot. Notice the reference to "bumpless" parameters in epicsInit.cmd; the saved data appears in the /log area of the IOC in use.

A master screen is provided, which can invoke both layers of the independent simulations.

6.12.3. Network Statistics

This Application adds a network statistics monitor capability to a system of one or more IOCs. For each IOC, the memory, cpu-usage, and Channel-Access client activity is monitored.

After building the STATS application itself, instructions are included to add the hooks to each primary application in each IOC. In the STATS dl subdirectory there is a master network mimic diagram which should be customized for your system. The generic screens do not need change. All the screens are accessible from the primary application areas.

For unusually heavy or light application loads, you may want to tailor the thresholds to provide the correct degree on sensitivity; this is done by changing the "warning" and "alarm" values in db/ioc.tmpl by using db/stats.db as a guide. (Hint: read the .DESC fields.)

SUMMARY

1. The Standard Applications Framework for EPICS (SAFE) provides an organized approach for building EPICS applications on a Unix host, include the cross-compiled VxWorks parts, the native Unix parts, and ancillary tools. SAFE provides disk directory layouts, scripts, recipes, and documents.
2. A particular feature of SAFE is the separation of the VxWorks/booting and IOC/initialization processes into two phases, each of which takes place in its own *area*, called *IOC* and *Application*, respectively. Each area has a creation script that builds and populates its directories, based on the SAFE configuration file and those from EPICS.
3. Although initially much installation activity takes place in the files associated with VxWorks and EPICS, except for updates these files remain quiescent. The Application and IOC areas refer to them with symbolic links. Most other development activity takes place in the IOC and Application areas, and all operational activity (actually "driving" the system) takes place in the Application area.
4. Sun C/C++ unbundled compiler: /usr/lang on SunOS and /opt/SUNWspro on Solaris. If you cannot obtain the Sun compiler, it is possible to copy binary versions of all of the EPICS Unix execut- ables from LBL to their installation directories, however, in some cases (such as Tcl/Tk), they will only run when placed in exactly the same path).
5. it is neither necessary nor desirable to have multiple copies of "extensions" when there are multiple copies of base; CONFIG_SITE in each release does the coupling. (Of course, a symbolic link will work as well.)
6. "base" will contain executables for all hosts that share the file system (for example, SunOS and Solaris—"sun4" and "solaris") and loadables for all targets for which you have BSPs. Do not checkout separate versions for this reason!
7. Only a subset of extensions are called out; with the exception of "BURT," these are just the ones listed in my Web page "Recommended EPICS documents." You can tailor the subset in config/CONFIG_EXTENSIONS.
8. The rec, dev, drv and dbd subdirectories are populated with files copied and slightly modified from Within the master EPICS area.
9. These four together perform two functions: "tailoring" each Application area to use just those existing records and devices/drivers needed; and "customizing" each Application by modifying or adding new records and devices/drivers, without affecting the master EPICS.
10. It is important to understand that this initialization activity is context-dependent, that is, it must be done from within the speci?c Application Area you are going to use. It refers to the symbolic links which bind this Application Area to a particular version of EPICS + VxWorks.

QUESTION WITH ANSWER

Q.1. What is SAFE?

Ans. The Standard Applications Framework for EPICS (SAFE) provides an organized approach for building EPICS applications on a Unix host, include the cross-compiled VxWorks parts, the native Unix parts, and ancillary tools. SAFE provides disk directory layouts, scripts, recipes, and documents. A particular feature of SAFE is the separation of the VxWorks/booting and IOC/initialization processes into two phases, each of which takes place in its own area, called IOC and Application, respectively. Each area has a creation script that builds and populates its directories, based on the SAFE configuration file and those from EPICS.

This document describes the procedures necessary for any project to use SAFE to develop and operate IOC applications. Although SAFE proper resides in the Application and IOC areas, its correct functioning is dependent upon some conventions on disk layout, on user/group naming, on network configuration, on installing VxWorks, and on installing EPICS.

Although initially much installation activity takes place in the files associated with VxWorks and EPICS, except for updates these files remain quiescent. The Application and IOC areas refer to them with symbolic links. Most other development activity takes place in the IOC and Application areas, and all operational activity (actually "driving" the system) takes place in the Application area.

Q.2. What is IOC Area?

Ans. The IOC area provides the following facilities:

- Linking to particular versions of EPICS and VxWorks is mostly driven from EPICS configuration files by means of a special script "iocMakeDir".
- The boot lines stored in the NVRAM of the VME board never change. An initial set is derived for each IOC based on configuration files.
- Each "logical" IOC-a combination of a particular board type plus its usage-has its own area.
- Switching between applications or combining several applications on one IOC is easily done.
- The VxWorks kernel can be customized for each IOC without altering the master installation, yet site-wide configurations may be shared.
- A per-IOC writable directory is easily accessible from applications which can thus use generic file names.
- Several IOCs can run same the same application relatively easily.
- Tailoring of network parameters that are common to all IOCs is kept in one place An initial set of commands is derived based on configuration files.

Q.3. What is Booting?

Ans. Here is the dynamic thread of the booting sequence.

- startup.cmd will be invoked after the kernel is loaded. It will:
  - tailor the console and login prompts;
  - call netInit.cmd for the common commands;
  - mount /log for all applications
  - call appinit.cmd to load the application(s)
  - ends.
- netInit.cmd:
  - performs the NFS mounts and other common commands;
  - returns.
- appinit.cmd
  - changes to the application subdirectory;
  - invokes the local, target-speci?c startup.<bsp>.cmd;
  - returns.
- startup.<bsp>.cmd:
  - invokes the local, target-speci?c epicsCode.<bsp>.cmd;
  - invokes the local, target-speci?c userCode.<bsp>.cmd;
  - invokes the local db.cmd;
  - invokes the local epicsInit.cmd;
  - invokes the local userInit.cmd;
  - returns.

In order to load multiple applications into one IOC, the generic application startup's must be broken apart (see Application area section).

Q.4. Compare VME vs. processor addresses.

Ans. The addresses found in module_types.h are true VME addresses, as set in the switches and jumpers of the I/O boards. From within the code running under VxWorks, however, processor addresses are used.

The relationship between these is set by hardware (such as the VMEchip2 ASIC of the MVME-167/177 series) which connects the CPU board's internal bus to the VME bus, and by software, particularly in a structure in sysLib.c called sysPhysMemDesc. For each IOC, you can modify this structure to arrange just the amount of processor address you want assigned to the A32 ("extended"), A24 ("stan-dard") and A15 ("short I/O") spaces of the VME bus.

In general, you should not change the A16 and A24 mappings at all. In a typical Motorola VME board,

A24 has the full 16 MBytes mapped from 0xf000'0000 to 0xf0ff'ffff, and A16 has the full 64KBytes mapped from 0xffff'0000 to 0xffff'ffff. Also, a small, 16Mbytes portion of A32 is mapped from 0x0200'0000 to 0x02ff'ffff. You can map much greater amounts than this (up to nearly 2 Gbytes), but doing so will consume large amounts of real RAM and take a long time to initialize. You may also change the caching. The VME space should always have caching disabled. By default,

WRS sets D32 accesses to A24 devices to be changed to two sequential D16 accesses; for some VME boards, this is just the wrong tactic and must be modified.

You should follow the convention that all EPICS drivers do, and never use absolute processor addresses; use the VME addresses and the function sysBusToLocalAddrs() to translate to processor address. The demonstration application "vmeProbe" is an excellent example of this; you are encouraged to use it for actual hardware testing and as a template for coding. (Notice careful use of the function vme- MemProbe() as well.) In general, I do not recommend attempting to access the VME bus directly from the VxWorks shell; often you will get a false, negative result merely because you have failed to properly translate to processor space or have not put enough "casts" in to generate the proper data width instructions (D8, D16, D32).

OBJECTIVE TYPE QUESTIONS

1. Your customer is unable to access Data Mover file systems by hostname, but is able to access them by IP address from HostA. HostB can access the Data Mover by hostname and IP address. What is the cause ?
   (a) lmhosts is empt (b) Duplicate IP address
   (c) Bad entry in HostA's hosts file (d) DNS reverse lookup entry is missing
2. Which protocol is designed to synchronize computer clocks over a network ?
   (a) FTP (b) NTP
   (c) CIFS (d) DHCP
3. Which command is used on a Sun host to obtain the Ethernet address of a network adapter ?
   (a) eth0 (b) NDD
   (c) ifconfig (d) ipconfig
4. Which command is used on a UNIX host to display all of the processes that are running ?
   (a) ps (b) tail
   (c) grep (d) egrep
5. A new level of EMCNAS code is now available. Where do you obtain the release notes ?
   (a) C4 Web Site (b) CCA Web Site
   (c) www.emc.com (d) Celerra Web Site

ANSWERS

1. (c) 2. (b) 3. (c) 4. (a) 5. (d)

☐☐☐

CHAPTER

7

KEY MANAGEMENT METRICS
(Thresholds, Availability, Capacity, Security and Performance)

7.0. INTRODUCTION

IT service organizations are increasingly implementing Quality and IT Service Management standards. The IT Infrastructure Library (ITIL, ISO20000), Control Objectives for Information and Related Technology (COBIT), Six Sigma, enhanced Telecom Operations Map (eTOM) and Sarbanes Oxley (SOX) are emerging as some of the more popular standards in addressing IT Service Management, governance, quality and operational issues.

This chapter considers the design and implementation of metrics in service organizations using one or more of the frameworks mentioned above. It uses the ITIL process structure and many principles from the ITIL and ISO20000 (originally produced in the UK as BS15000) as a basis.

It is a general guide to the use of metrics as a mechanism to control and steer IT service organizations. Implementing IT Service Management as a series of interlocking processes - with 'process' strictly defined—enables a consistent view to be taken across the many disciplines that exist in a modern IT department.

This consistent view has been adopted as Best Practice by thousands of organizations across the

Inside this Chapter

world, with excellent results. The ITSMF is an independent organization that champions the cause of IT Service Management in many countries by holding events, working to improve the practice advocated by ITIL through knowledge and experience sharing, whereby driving continuous refinement of the practice, and producing chapters such as this.

The ITIL processes all devote a section to possible metrics, giving an excellent starting framework for setting up metrics. For example, there is a chapter focusing on Key Performance Indicators (KPIs) in the ITIL Best Practice chapter 'Planning to Implement Service Management'.

This guide deals specifically with the issue of setting up metrics in the context of IT Service Management frameworks, with a special focus on ITIL.

A major reason for writing this chapter is that many organizations have found it very difficult to use metrics properly. This chapter will deal with the causes of the difficulties to implementing metrics and will present workable solutions.

The chapter is a general guide to the design, implementation and use of metrics as a mechanism to control and steer IT service organizations. It also provides specific recommendations for applying metrics across the ITIL, ISO20000 and other processes, discussing the rationale of the recommendations. This enables an organization to implement the metrics as described directly as a first-pass solution that can be benchmarked against other organizations. But they can also be used as a starting point for customizing particular metrics.

Badly designed metrics can be actively harmful to an organization's proper functioning. Producing a set of metrics that avoids the pitfalls and delivers genuine value is not easy. This chapter will make that task much simpler and less error prone.

The recommended audience for this chapter is service managers, process owners, consultants, general IT management and anybody interested in mastering metrics of IT Service Management.

A worldwide review team of matter experts who have provided the benefit of many decades of collective IT and Service Management experience has extensively reviewed this chapter. You can rely on it as a guide to Best Practice in the field.

7.1. THRESHOLDS

A 'metric' is just another term for a measure. In IT, metrics have come to mean particular things to different people. Though this chapter is about metrics, it is important to remember why we use metrics. Simply measuring things for the sake of it is expensive and pointless. Metrics themselves are not an end. Metrics are an important part of the Management System that steers and controls IT in the desired direction. As we will see, Metrics must be designed in line with customer requirements, they must be benchmarked to ensure that they are achievable and they must be monitored to ensure that they keep within desired thresholds with action taken to correct any problems. They also are the target of the Continuous Service Improvement Programme (SIP), as processes and services are continuously improved, so are the metrics that measure them. It is important to understand what the business' objectives are and ultimately arrange that all measuring, monitoring and control is aligned to attaining these objectives. This chapter discusses ways to ensure this.

7.1.1. Metrics as Management Information

The business has to understand how well its business processes are performing. IT plays an important role in three ways to assist with this.

Firstly, IT nowadays often is responsible for providing accounting, logistical and other direct services to business processes. Accordingly, the measures are provided in management reports to the various business units—for example SAP reports on sales, provided to the Sales team.

Secondly, IT provides services to business processes documented in the Service Catalogue, with the detail of delivery defined in the various SLAs set up between the Service Level Manager and business customers. These measures are shown as exception reports to Key Performance Indicators (KPIs) based on negotiated SLAs. Trends in these reports show how IT is improving its ability to provide services to a high standard.

Thirdly, IT itself is an important business process, a business contingency plan without a substantial section on IT service continuity is rare! Thus, IT reports on the operation of IT processes are part of the management information required by the business. These metrics form the bulk of the discussion in this chapter. They're based on the measurement of IT Service Management process operation.

This information is, however, business information, so detailed technical metrics are not appropriate. Rather the condensed results of measurement of IT processes can be presented in business terms. Ultimately business measures are monetary measures so the aim of IT is eventually to provide Return on Investment (ROI), Return on Capital Employed (ROCE), Economic Value Added (EVA) or any other expression.

Before this can be done, however, IT Financial Management, in ITIL terms, must have reached a high level of maturity. Until then, measures of process efficiency, along with KPIs against SLAs provide the most complete picture of IT's services to the business.

Once clear and relevant metrics have been designed, it is important that they are presented clearly. Various ways of achieving this are discussed below, such as 'balanced scorecard', 'traffic light' and 'dashboard' systems. Different reporting methods and different sets of metrics will be appropriate for different audiences. The naming of metrics is also important so that, if a metric is changed from one reporting period to another this is made clear.

If processes are not implemented in a consistent, repeatable manner, the metrics produced by them will be unreliable. It is important, as ISO20000 emphasizes, for there to be a sound Management System in place with sufficient maturity and process management to be part of an organization's way of doing things for metrics to provide useful measurement.

7.1.2. Metrics for Management Control

When something in business is measured, particularly when this measure is made the responsibility of a manager or a team of people, the behavior of the people measured changes.

If the metrics are well designed and the objectives of the metrics are in line with business requirements, this behavior will tend to be in line with these business requirements. In other words, a well designed and measured metric is a method of control. If a metric is not well designed, is not in line with actual business requirements or is not measured correctly then this 'control' can drive behavior in the opposite direction and harm the operation of the business.

> There are many examples of this. Just one will suffice to show the problem. In the UK, the government decided that the length of time patients had to wait for operations was a useful metric. To measure this it set targets to reduce the length of waiting lists. The result was exactly the behavior the government had requested. The waiting lists got shorter and the targets were met. However patients were actually waiting just as long for their operations as before. Hospital administrators were not allowing patients to be entered onto the waiting list until their operation had been scheduled to a time within the target waiting time. Thus, a second, informal, waiting list came into being. On it were those waiting to be allowed onto the waiting list!.

This example is a good one for a few reasons. One is that it shows the ingenuity of people in working to produce exactly what is asked for, and measured, rather than what is actually sought, the *letter* of the law rather than the *spirit* of the law. It also shows that a measure must be designed to actually measure what is important. In this case a waiting list was measured. However, it would have been more appropriate, but quite possible more difficult, to measure patient referral dates against patient operation dates. So, the design of the measure is important. The final problem was that the process itself was not measured, so when a new process (the informal waiting list for the waiting list) was invented by hospital administrators, it was not visible to the government auditors.

Another problem many organizations have met is indecision. If metrics have been badly designed, they have to be changed, of course. However, if they are changed to try to address the immediate problem, rather than carefully designed, the new metrics may prove almost as bad. This will then mean that they have to be changed again. If metrics are changed every few months there is no reason for anybody to work hard to achieve them. People soon work out that they only have to wait for the change, and there will be a new period of uncertainty in which there is no effective metric and hence no effective management control. An organization under the effect of constantly changing metrics is likely to be worse off than one with no metrics at all.

We are all happy to work for goals that we think have a reasonable chance of attaining. We also work better if we are given recognition, and sometimes even praise, when we do a job well. If we are asked to do an impossible job, we know that we can only be blamed for not managing to do it. The natural human response to this is to give up and simply go through the motions of trying.

Metrics must be set to be perceived as achievable and make business sense so as to encourage positive change of behavior.

The moral of this is that metrics must be achievable and recognition and praise must be given to those who achieve and excel in over-achieving against the metrics.

The various process metrics enable the IT organization as a whole to measure the effectiveness of managers in implementing, improving and maintaining the quality of the delivery in their areas of responsibility.

7.1.3. Metric integration and Reporting

Metrics do involve detailed measures of technical matters. This is unavoidable. However, once the key metrics have been set, these can be reported using the 'traffic light' or 'dashboard' methods, discussed in detail in chapter 7. This method allows a 'drill-down' into the detail to occur when a problem is isolated at the top level.

If the metrics for different processes are designed in a similar way, the management of these processes can be compared with each other. Though the Change Management

Process is a very different thing from the Availability Management Process the management, maturity and effectiveness of these two processes can be compared by having similarly structured integrated metrics, particularly if these metrics include an accurate measure of customer satisfaction, in which the 'customer' is defined as the main beneficiary from the output of the process.

For example: if Availability Management has a score of 8.5 on the weighted sum of its ten metrics and Change Management has a score of 10.8 on its metrics, we can see that Availability Management is less in control of its processes than Change Management, even though the processes themselves are quite different.

7.1.4. Metrics Aligned to Stakeholders

Communication is a vital part of IT Service Management. If stakeholders are being informed by metrics, they can contribute to the success of the enterprise by supporting the openness and transparency, and by seeing the improvement. Each section below discusses the needs of particular stakeholders and how they fit into the customer relationship diagram (Fig. 6.1).

This enables appropriate communication of metrics, their results and what these results mean in terms of delivery of services to stakeholders. Stakeholders need to be an integral part of IT definition and satisfied with the results they see from their active involvement.

Communication is a two-way process so it is important to integrate requirements from the various stakeholders and use their involvement throughout to improve service delivery and process operation.

For this to be handled properly, it is important that it is dealt with using a carefully constructed communication plan, using stakeholders to assist in its construction and review.

Customer

In this chapter, by using the word 'customer', we usually mean the purchaser of a service, be it internal or external. In this section we refer to the end customer of the business. Ultimately all business effort ought to be directed towards the end customer. IT provides a support function for the business processes, but sometimes the IT contribution is experienced by the business, end customer (the ultimate customer) as well.

In these cases, to be sure that we have contributed effectively to the business, the business must measure the satisfaction of all end customers. This measurement must, however, be handled sensitively. If we demand surveys too frequently from customers, this activity itself will become a negative factor to them. If we don't solicit advice on how well we are doing often enough, there is a danger of not being aware of service incidents for long enough for them to impact our customer satisfaction seriously.

User

Since users do not negotiate the terms of their service and do not pay for it, they are not direct customers of IT. However, as stakeholders, their satisfaction with the service is vital. If users are not satisfied as they are being supplied with poor quality services, eventually our customers will not be satisfied.

Employee

Employees within IT have the responsibility to deliver service processes. IT, in return, has the responsibility to supply good working conditions with fair evaluation and reward. If employees are not satisfied the service levels provided will be less than they ought to be, no matter how well defined the metrics are. Thus, it is an imperative of IT management to measure staff morale and be sensitive to changes that might impact it negatively.

Employees are happier when they feel recognized as a stakeholder. Also, they need to be able to identify with their employer, out of positive respect for the business, and approval of the processes. Communicating metrics to employees clearly and openly, enables them to understand where IT is doing well and where there are issues that require further effort. Effective communication enables employees to address issues and to find an inspiration to further effort.

Board

Senior management and the Board have a particular need, as stakeholders. In order for the business to strive, they require advance warning of any potentially serious incidents so that urgent corrective actions can be taken. An open and transparent communication with the Board can help as these incidents can be identified before they become too serious to remedy.

Good news is also important to the Board. It can be communicated both within and outside the company to enhance its reputation. The attainment of ISO20000, for example, ought to be an opportunity for the Board to report the achievement with pride.

Other Stakeholders like Government and Shareholders

Though these stakeholders are important to the business, the correct communication to them is the Board's responsibility. If IT has messages that can reassure or warn such stakeholders it is important that they are communicated first to the Board. In this way, the Board can decide on the appropriate channel and method of communication.

7.2. AVAILABILITY

The agreement between the business organization, the customer, and IT is negotiated by the Service Level Manager and results in a set of defined Service Level Agreements (SLAs). These SLAs define what service levels IT agrees to provide.

These SLAs are used by the Service Level Manager to define the Operational Level Agreements (OLAs) and, where third parties are involved, the Underpinning Contracts (UC) that enable the service to be delivered.

The Critical Success Factors (CSFs) for IT are defined by the OLAs - if you look at it from a bottom-up point of view. If an OLA is met then the relevant CSF will be satisfied - if the match between them is good. Actually OLAs are derived from SLAs which rely on CSFs, so are broader in scope than a particular OLA may be. For a member of the organization, though, the CSF can be seen as the goal of the OLA.

CSFs are the service delivery measures that must be met to satisfy the SLAs. Each CSF can then be used to define a Key Performance Indicator (KPI) that is a measure of whether the CSF is being delivered.

Thus the Entire Chain

Customer Requirement > SLA > OLA/UC > CSF > KPI > Monthly Report

is driven directly by the customer requirements and the KPIs can be measured and reported back to the customer to show how effectively the IT organization is meeting the agreed service levels.

7.3. CAPACITY AND SECURITY

As we see above, the KPI provides a customer facing metric that measures the success with the SLAs defined with the customer.

This enables IT management to know each month whether it is doing well or not. This is too late to do anything about it! It is no good having a petrol gauge in your car that tells you that the tank is empty. You need one to tell you how full the tank is so that you can fill up before it becomes a problem.

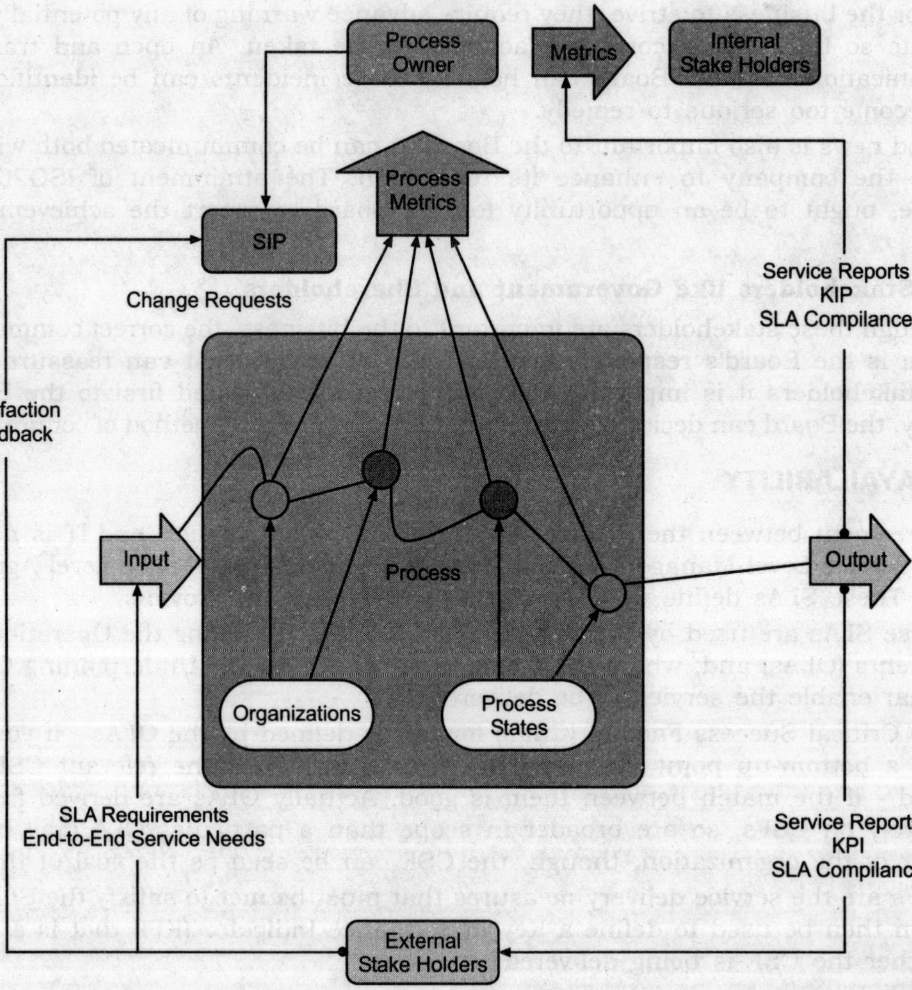

Fig. 7.1. General Process Schematic

In just the same way, IT management must have measures that show how the organization is operating daily or weekly so that corrections can be made before SLAs are compromized.

This is where process metrics come in to play. The ITIL approach to IT Service Management defines delivery in terms of services to customers. This is why the KPI is the proper measure of service delivery. However, ITIL also defines the operation of the various parts of IT in terms of processes. Each of these processes can be seen as an engine that takes certain input and processes it into output, as Fig. 7.1 points out.

The process flow between Organizations and/or Process States is shown along with the contribution to Metrics. This diagram also shows the alignment between this process and both internal and external stakeholders. Dark gray represents processes and organizations internal to IT whilst *pale gray* represents external connections and organizations.

As can be seen, for the process owner to control the working of the process to deliver its goals, KPIs are defined for the process. These are measured daily or hourly and triggers on thresholds decide escalations to enable corrective actions.

Each process, however, working with the other ITIL processes, must be managed by metrics that can be reported to IT management and to other stakeholders, to contrast the process operation of the different IT processes. The careful selection and measurement of the process metrics as shown enables the management of the process, as a process, the measurement of the process owner and, where relevant, the process team. These metrics include some of the KPIs used by the process owner. He, however, is likely to use a larger set of metrics, giving more detail of day-to-day changes in process delivery, thus providing more hands-on management.

7.4. PERFORMANCE

Over a longer period of time it is possible to measure previous results of metrics and compare one month or year to previous months and years. These comparisons can be used to set performance improvement goals.

There are two problems with this. Firstly, when metrics are introduced, it is not clear what an acceptable or ideal level of performance might be. In order to establish this, it is necessary to produce an initial base level or 'benchmark'. This is often done by running the metrics for two or three periods and then taking the values produced as a 'benchmark' of the minimum requirement and working from there.

The second problem is knowing whether this benchmark compares well or badly with other organizations. If your metrics are uniquely set for your organization this will be an impossible question to answer! Nobody else uses your metrics, so only you know what values are possible.

There are two approaches to resolving this. Some research organizations have produced lists of standard metrics with average results across a number of organizations. If you measure these same standard metrics then these figures can be used to identify how your organization sits relative to the average.

The other approach is to use either a standard set of metrics, or a set modified from a standard set, such as those in this chapter. Then it is possible to compare your metrics directly with other organizations using the same, or very similar, metrics.

A hybrid approach is also possible. If you implement your own metrics, but also measure two or three of the published research metrics, then benchmarking the research metrics can give an external view, giving an insight into how well the processes work.

If, for example, the research metrics show your organization to be between 95 per cent and 115 per cent against the average, then the other metric goals can be scaled to fit within the same range. At some future time, if your own metrics are, say, 120 per cent to 130 against this benchmark then it can be reasonably assumed that you are at approximately that level relative to the organizations in the benchmark.

Using the method presented in this chapter of a reasonably standard set of metrics across all processes allows a third approach. It is possible to benchmark your processes against each other. ITIL is usually implemented in phases, so, if your weighted goals for a well established process, say Change Management, are achieved with an average of 130 per cent, then the benchmark goals for a new process can be set initially to give a weighted average result of, say, 80 per cent.

This way process improvement over time can be measured to see how long this new process takes to equal the level of Change Management.

7.5. WHY METRICS?

- *Cheshire Puss: she began, rather timidly, as she did not at all know whether it would like the name: however, it only grinned a little wider.*
- *Come, it's pleased so far–thought Alice, and she went on. Would you tell me, please, which way I ought to go from here?*
- *That depends a good deal on where you want to go to–said the Cat.*
- *I don't much care where–said Alice.*
- *Then it doesn't matter which way you go–said the Cat.*
- *So long as I get somewhere–Alice added as an explanation.*
- *Oh, you're sure to do that–said the Cat–if you only walk long enough.*

Alice in Wonderland, Lewis Carroll

We rely on metrics every day for many of the things we do. We simply do not think of them in quite that way! It is important to put metrics into perspective because there is a danger of being carried away. If you know what you want to do, or where you need to go, then you can decide what you need to measure to know how you are doing your way to get there and how it will help you. If you are given a machine with many different instruments, there is a temptation to find the instruments themselves exciting rather than what that machine is really useful for.

As with Alice, if we do not care where we are going it does not matter what we measure. If we do not measure where we are going then we are sure to get somewhere. But probably not where we would like to be!

7.5.1. Metrics as an Instrument

The speedometer in a car or aircraft generates metrics; it provides a measure of how fast the driver is going. He will be aware when he's driving too fast, because he knows the maximum speed, and therefore knows what to do: brake or release the gas.

Before the days of very fast cars and speed traps, this was an unnecessary thing to measure. After all, you could tell how fast you were going from the wind in your hair and your speed around corners!

From even the earliest days a much more important instrument was the fuel gauge, particularly in aeroplanes. If you run out of fuel in a car you either have a long walk or, if you have come prepared, have to fill up from a spare can (a dangerous option).

When you are flying your options are more limited so an accurate fuel gauge is that much more important.

In the language of metrics, having enough fuel to complete your journey is a Critical Success Factor (CSF) and the instruments, or metrics, used to measure it are your Key Performance Indicators (KPIs). If you are to concentrate on your driving or flying you need to keep these to a reasonably small number. So, as we have said, a fuel gauge is probably your top KPI.

An oil level indicator will be a close second since this can cause your engine to seize. These days, however, though these instruments are important, they are usually so reliable that we only need to know when it is time to fill up or when there is danger of overheating.

With these metrics, it is quite easy to see how the process model works. Consider the speedometer. The input will come from a transducer on the wheel. The output is the level of the indicator needle on the dial. The throughput, or the operation of the process itself, is the gearing (or now electronics) in the speedometer itself that translates the transducer input in so many thousands of turns into the output in so many kilometers per hour. The appropriate target levels of outcome will vary.

On a motorway they will be different from on a farm track. And the optimal outcome for the engine, as shown by the speedometer and the rev counter (tachometer), may not be appropriate because of legal speed restrictions.

The speedometer and odometer are less crucially important, to the functioning of the vehicle, but are more important to our trip. If we know the distance and at what time we are due to arrive, we can adjust our speed depending on the distance we see that we have come. If we have plenty of time in hand, we can even take breaks to ensure that we are alert enough to drive safely.

With all this in hand, our KPI, the matter that will decide if we arrive on time or not, might well be our ability to read the map properly, or a cell phone to help us navigate the last few kilometers to our destination with the help of landmarks and advice from those we are visiting.

These are not instrumented (unless we have GPS in the car!) so it is less easy to measure unless things go wrong. This last point is another one to keep in mind with IT metrics!

7.5.2. Metrics as a Control

The true control of a car is through the steering, the accelerator, brakes and gears. These allow us to control the car through minute changes to the road surface, unexpected hazards and other short-term contingencies. These are similar to day-to-day management in a company–or what is sometimes called 'micro-management', too much concern with the fine detail. This is not to say that the fine detail is not important, if you run over a log in the road and have a puncture, your journey time will certainly suffer!

The real long-term control comes from good planning, just like anything else! Properly selected metrics, with proper management, encourage properly defined activities that allow us to increase the certainty that we will arrive on time at the end of the journey. How often the car is serviced is one very long-term but important metric.

Good planning and map reading ensure that the journey does not have to be carried out at top speed with speeding, fast cornering and severe braking. All that will enable the journey to complete, but will also increase risk and make the journey uncomfortable for the passengers.

They might decide to go by train next time!

Some instruments, the rev counter, for example, do not seem immediately valuable, some cars don't even have them. A sensitive driver, however, can use one to get the optimal performance from the car, without putting the engine under too much strain. If you were running a rally, points could be given to competitors who never redline the rev counter, but complete in the fastest time with the use of the minimum fuel. That calls for excellent driving, excellent planning and excellent navigation.

This long analogy applies pretty directly to IT metrics. In IT, there are KPIs that are so important that we just expect them to be done, like the fuel gauge being accurate! Other metrics are difficult to gather and only very sophisticated and mature IT organizations will need them to get optimal performance, like rev counters. Meanwhile most IT organizations will benefit greatly from using metrics to control the ride so that life is not all fire-fighting, sharp cornering, extreme braking, and the business, and people in IT, are comfortable enough to give of their best without unnecessary stress. It is important to remember that the measure is not enough to ensure the result - the activity must be managed in line with the metric - a well designed speedometer will not make you obey the speed limit!

Five different methods of coordinating organizational activities have been identified (Mintzberg). These can be likened to the various instrument options in the analogy, as Table 7.1 shows.

Table 7.1. Five methods of coordinating organizational activities and their analogies to driving a car (after Mintzberg)

Method of coordination	Analogy to driving a car
Direct Supervision	The supervisor instructs the driver to get the package to the customer by 17:00 this evening.
Mutual Adjustment	The drivers plan their breaks so that notall busses are unavailable at the same time.
Standardization of working practices	Drivers are taught to follow a predesigned process.
Standardization of output	Drivers arrive at their destination with intime windows of ten minutes.
Standardization of capabilities	Drivers have advanced driving certificate.

In the above examples, we see various ways of requesting desired results. Some require underlying metrics to be in place to help the measure—checking that all drivers have advanced driving certificates, for example. Others are ad hoc and need to be documented individually.

For IT management we can see that much of what occurs today is at the 'Direct Supervision' level. With the rise of ITIL, it is possible to put in place metrics that allow ITIL qualified people to implement standardized processes with standardized metrics. ISO20000 makes it clear that this works only if it is part of a coherent Management System.

7.5.3. Metrics and Innovation

Metrics can only work if there is a process in place to be measured. Measuring how things happen without a process allows no consistent method of improvement, nor even a coherent way of discovering what prevented the metric occurring or how.

Much criticism of metrics comes from people who are opposed to processes rather than to metrics themselves. This is discussed more on resistance. It is worth noting here that it is easy to agree to work according to a particular process and then carry on doing things just as before. Unless there are appropriate process metrics that will soon show that the process is not working.

Organizations that believe they have implemented processes but find that they appear to give little benefit may discover nothing has really changed when process metrics are established. For this reason, as well as the ones already mentioned, it makes sense to implement metrics at the same time as the processes are piloted, but this has often not been done!

There are a number of reasons for objecting to processes, and hence to metrics. One is simply that they are a change and most of us do not welcome change. Another is that they make a job boring as they force it to be done in a particular way each time. A further objection is that the measurement of process metrics exposes people to criticism and may endanger their jobs.

There is an element of truth to all the above objections, particularly if the reason for processes is not well communicated.

An analogy might help here. Riding a bicycle is a process. Once we learn how to do it, and those that have remember the wobbly start, the fear of falling off and the scraped knees that accompanied the learning, we no longer think about the detail. In fact, it has been pointed out that, if you think too carefully about what you are doing when riding a bicycle, you are likely to fall off! The important thing is that we then can concentrate on where we are going, what the view is like and the pleasure of being outdoors. If we never turn riding a bicycle into a wellknown process, we will condemn ourselves to always finding it difficult, falling off and not enjoying the journey. We probably will not complete many journeys and certainly not many long ones.

So, it is with IT service management or business processes. Once they are in place and working then it is possible to concentrate on the more interesting matters of introducing new things and fire fighting ceases to be a way of life. In an organization with good processes, people are free to innovate and give of their best in a calmer, more relaxed environment than in an organization that moves from crisis-to-crisis.

Yes, metrics do make problems visible and they have to be fixed. However, the object of the exercise is not to blame individuals, but to fix the process. It takes time for people to realize this, but it is vital for processes and their metrics to become accepted.

If metrics provide only the Red, Amber and Green traffic light reporting system, managers are tempted to look only for the Red metrics to see what the problems are. Reporting a Gold level where metrics are being not only met, but exceeded, is one possible way of enabling more positive celebration to happen, not just fault finding ! It is, again, important for all to know, and for it to be emphasized, that process metrics are designed to enable processes to be improved, not find fault with people !.

7.5.4. Costs

Measuring can cost a lot of money. One service management consultant was happy to save a company millions of pounds by closing down a department of over a hundred people who did nothing but produce reports. He found out all these reports were easier and cheaper to obtain from elsewhere in the organization.

One international organization sent out a monthly chapter of reports, called the 'Blue Chapter', because of its cover. One month, to test its importance, the management deliberately decided not to send out the report but wait for requests for it. Only one came. They were pleased to find that one person, out of the twenty who received the very expensive report, was interested in having it and asked why he wanted it. "My four year old daughter loves it because she can draw on the back of each page and it is very good paper", was his reply.

Reports are only worthwhile if they are used and they will only be used if they tell us something interesting. Metrics must be designed to measure important matters and reported in a clear and simple manner. For most managers, a one-page summary on a web site will be perfectly suitable to their needs, as long as any potential problems are communicated clearly. Hiding bad news in the notes at the back of reports might work for publicly listed companies, but it is not a sensible approach for IT !

Metrics that report on costs are important and, in the IT Service Management scheme, the responsibility of Financial Management. The level, or maturity, of the organization will determine at what level of cost metrics are produced. It is important not to report costs inaccurately. Waiting until a proper costing model has been agreed with the business by the Financial Management function through the SLA process would be better.

7.5.5. Benefits

This chapter ought to have made most of the benefits clear. To summarize:

- Metrics provide the instrumentation necessary to control an organization.
- Metrics make it easier to concentrate on the important matters.
- Well-presented metrics make it easy to spot danger in time to correct it.
- Metrics can improve morale in an organization.
- Metrics can stimulate healthy competition between process owners.
- Metrics help align IT with business goals.

7.6. WHERE TO USE METRICS

IT service management software tools provide many possible metrics, most of which will be interesting to people in the relevant departments or running the relevant processes. If only these metrics are relied upon then the 'tail wags the dog'. The organization knows what the designer of the software thought it should know, not what business IT alignment demands.

For this reason it is important to start at the other end, design the processes for your organization, then design the metrics to measure them. After this, it is possible to decide how to implement them in software.

Since metrics are designed to allow an organization to control its destiny, somebody must be responsible for them. If two people are responsible for the same metric, neither person is as the actions of one could prevent the other from achieving the desired result.

With process driven IT Service Management, whether you're using ITIL, Six Sigma, COBIT or all three, it is important to distinguish between individuals, organizations or departments, and processes. For a metric to be a successful control, it must be the responsibility of somebody, an individual, who has the authority necessary to influence the outcome of the metric. There is no point, as King Canute showed, of giving somebody responsibility for the tides in the sea, unless that person also is able to influence them!

For each metric, it is necessary to establish who the responsible person will be. Will it be a departmental manager or will it be a designated process owner who might not have any direct reports.

7.6.1. Departments

In IT service management, the most obvious two departmental entities that every organization has are the Service Desk (formerly known as the Help Desk) and the ICT Infrastructure Management team (formerly known as Operations). These two departments will have department heads responsible for metrics measuring their departmental efficiency, and for the processes for which their departments are responsible.

The service desk is often responsible for the incident management process. The manager of this department will then be the process owner responsible for the metrics.

Depending on the size and structure of the organization, there might be other departments in IT. These either carry ownership of one or more ITIL processes or provide services to these processes. It is important that appropriate metrics are established, in line with the top level processes but applying to the department's manager.

Note: For example, consider an administrative department that provides reporting, of both ITIL process metrics and SLA compliance. It may have other duties as well, but metrics that test the accuracy and timeliness of production of these reports must be in place for the manager of this department. Such a department may also well have responsibility for the communication plan, an important plan that provides services to all other processes and must have its own measures of success.

7.6.2. Processes

Many processes don't have their own department to run the process, but rather run across many departments and organizations to deliver the end-to-end service. The person responsible for these is the process owner who must have authority to ensure that the process metrics are met, even with only influence management authority.

Note: A good example of this is the Change Manager who is responsible for the Change Management process. One of the most important requirements for the implementation of IT Service Management along ITIL quality lines is an effective corporate sponsor. Change Management is always going to involve influence management, as so many different parties are involved. The Change Manager is responsible for final approval of changes, under advice. This is not always going to be a popular decision and there are opportunities for political, rather than business, considerations to come into play. The corporate sponsor is needed to provide the required support to the influence management of the Change Manager. This is possible if there is evidence that the Change Manager is doing a good job. Without metrics to prove it, this can be a matter of opinion and political undermining of the Change Manager becomes easier. Well chosen metrics for the Change Management process can help identifying organizational areas that are not as cooperative as they could be. Working with them to iron out any process difficulties can be made considerably easier by a strong executive sponsor.

It is worth repeating one point. People are very sensitive to criticism, real or imagined. Poorly performing processes, identified by metrics are an opportunity for improvement. They are not supposed to be an attack on people! It is important that this point is emphasized in communications.

SUMMARY

1. A 'metric' is just another term for a measure. In IT, metrics have come to mean particular things to different people. Though this chapter is about metrics, it is important to remember why we use metrics. Simply measuring things for the sake of it is expensive and pointless.
2. Metrics themselves are not an end. Metrics are an important part of the Management System that steers and controls IT in the desired direction.
3. Metrics must be designed in line with customer requirements, they must be benchmarked to ensure that they are achievable and they must be monitored to ensure that they keep within desired thresholds with action taken to correct any problems. They also are the target of the Continuous Service Improvement Programe (SIP), as processes and services are continuously improved, so are the metrics that measure them.
4. It is important to understand what the business' objectives are and ultimately arrange that all measuring, monitoring and control is aligned to attaining these objectives.
5. Firstly, IT nowadays often is responsible for providing accounting, logistical and other direct services to business processes. Accordingly, the measures are provided in management reports to the various business units - for example SAP reports on sales, provided to the sales team.
6. Secondly, IT provides services to business processes documented in the Service Catalogue, with the detail of delivery defined in the various SLAs set up between the Service Level Manager and business customers. These measures are shown as exception reports to Key Performance Indicators (KPIs) based on negotiated SLAs. Trends in these reports show how IT is improving its ability to provide services to a high standard.
7. Thirdly, IT itself is an important business process, a business contingency plan without a substantial section on IT service continuity is rare! Thus, IT reports on the operation of IT processes are part of the management information required by the business. These metrics form the bulk of the discussion in this chapter. They're based on the measurement of IT Service Management process operation.
8. Metrics do involve detailed measures of technical matters. This is unavoidable. However, once the key metrics have been set, these can be reported using the 'traffic light' or 'dashboard' methods, discussed in detail in chapter 7. This method allows a 'drill-down' into the detail to occur when a problem is isolated at the top level.
9. If the metrics for different processes are designed in a similar way, the management of these processes can be compared with each other. Though the Change Management Process is a very different thing from the Availability Management Process the management, maturity and effectiveness of these two processes can be compared by having similarly structured integrated metrics, particularly if these metrics include an accurate measure of customer satisfaction, in which the 'customer' is defined as the main beneficiary from the output of the process.
10. Since users do not negotiate the terms of their service and do not pay for it, they are not direct customers of IT. However, as stakeholders, their satisfaction with the service is vital. If users are not satisfied as they are being supplied with poor quality services, eventually our customers will not be satisfied.

QUESTION WITH ANSWER

Q.1. What do you mean by Thresholds of metrics?

Ans. A 'metric' is just another term for a measure. In IT, metrics have come to mean particular things to different people. Though this chapter is about metrics, it is important to remember why we use metrics. Simply measuring things for the sake of it is expensive and pointless.

Metrics themselves are not an end. Metrics are an important part of the Management System that steers and controls IT in the desired direction. As we will see, Metrics must be designed in line with customer requirements, they must be benchmarked to ensure that they are achievable and they must be monitored to ensure that they keep within desired thresholds with action taken to correct any problems. They also are the target of the Continuous Service Improvement Programme (SIP), as processes and services are continuously improved, so are the metrics that measure them.

It is important to understand what the business' objectives are and ultimately arrange that all measuring, monitoring and control is aligned to attaining these objectives. This chapter discusses ways to ensure this.

Q.2. Define that the Metrics as Management Information

Ans. The business has to understand how well its business processes are performing. IT plays an important role in three ways to assist with this.

Firstly, IT nowadays often is responsible for providing accounting, logistical and other direct services to business processes. Accordingly, the measures are provided in management reports to the various business units—for example SAP reports on sales, provided to the Sales team.

Secondly, IT provides services to business processes documented in the Service Catalogue, with the detail of delivery defined in the various SLAs set up between the Service Level Manager and business customers. These measures are shown as exception reports to Key Performance Indicators (KPIs) based on negotiated SLAs. Trends in these reports show how IT is improving its ability to provide services to a high standard.

Thirdly, IT itself is an important business process, a business contingency plan without a substantial section on IT service continuity is rare! Thus, IT reports on the operation of IT processes are part of the management information required by the business. These metrics form the bulk of the discussion in this chapter. They're based on the measurement of IT Service Management process operation.

Q.3. How can Metrics be used for Management Control?

Ans. When something in business is measured, particularly when this measure is made the responsibility of a manager or a team of people, the behavior of the people measured changes.

If the metrics are well designed and the objectives of the metrics are in line with business requirements, this behavior will tend to be in line with these business requirements. In other words, a well designed and measured metric is a method of control. If a metric is not well designed, is not in line with actual business requirements or is not measured correctly then this 'control' can drive behavior in the opposite direction and harm the operation of the business.

There are many examples of this. Just one will suffice to show the problem. In the UK, the government decided that the length of time patients had to wait for operations was a useful metric. To measure this it set targets to reduce the length of waiting lists. The result was exactly the behavior the government had requested. The waiting lists got shorter and the targets were met. However patients were actually waiting just as long for their operations as before. Hospital administrators were not allowing patients to be entered onto the waiting list until their operation had been scheduled to a time within the target waiting time. Thus, a second, informal, waiting list came into being. On it were those waiting to be allowed onto the waiting list!

Q.4. Define Metric integration and Reporting.

Ans. Metrics do involve detailed measures of technical matters. This is unavoidable. However, once the key metrics have been set, these can be reported using the 'traffic light' or 'dashboard' methods, discussed in detail in chapter 7. This method allows a 'drill-down' into the detail to occur when a problem is isolated at the top level.

If the metrics for different processes are designed in a similar way, the management of these processes can be compared with each other. Though the change management process is a very different thing from the availability management process the management, maturity and effectiveness of these two processes can be compared by having similarly structured integrated metrics, particularly if these metrics include an accurate measure of customer satisfaction, in which the 'customer' is defined as the main beneficiary from the output of the process.

For example: if availability management has a score of 8.5 on the weighted sum of its ten metrics and change management has a score of 10.8 on its metrics, we can see that availability management is less in control of its processes than change management, even though the processes themselves are quite different.

Q.5. What do you mean by availability of metrics?

Ans. The agreement between the business organization, the customer, and IT is negotiated by the service level manager and results in a set of defined service level agreements (SLAs). These SLAs define what service levels IT agrees to provide.

These SLAs are used by the service level manager to define the operational level agreements (OLAs) and, where third parties are involved, the Underpinning Contracts (UC) that enable the service to be delivered.

The critical success factors (CSFs) for IT are defined by the OLAs—if you look at it from a bottom-up point of view. If an OLA is met then the relevant CSF will be satisfied—if the match between them is good. Actually OLAs are derived from SLAs which rely on CSFs, so are broader in scope than a particular OLA may be. For a member of the organization, though, the CSF can be seen as the goal of the OLA.

CSFs are the service delivery measures that must be met to satisfy the SLAs. Each CSF can then be used to define a Key Performance Indicator (KPI) that is a measure of whether the CSF is being delivered.

Thus the entire chain:

Customer Requirement > SLA > OLA/UC > CSF > KPI > Monthly Report

is driven directly by the customer requirements and the KPIs can be measured and reported back to the customer to show how effectively the IT organization is meeting the agreed service levels.

Q.6. Define capacity and security?

Ans. As we see above, the KPI provides a customer facing metric that measures the success with the SLAs defined with the customer.

This enables IT management to know each month whether it is doing well or not. This is too late to do anything about it! It is no good having a petrol gauge in your car that tells you that the tank is empty. You need one to tell you how full the tank is so that you can fill up before it becomes a problem.

In just the same way, IT management must have measures that show how the organization is operating daily or weekly so that corrections can be made before SLAs are compromized.

This is where process metrics come in to play. The ITIL approach to IT Service Management defines delivery in terms of services to customers. This is why the KPI is the proper measure of service delivery. However, ITIL also defines the operation of the various parts of IT in terms of processes. Each of these processes can be seen as an engine that takes certain input and processes it into output, as Fig. 7.2 points out.

The process flow between organizations and/or process states is shown along with the contribution to Metrics. This diagram also shows the alignment between this process and both internal and external stakeholders. Dark gray represents processes and organizations internal to IT whilst pale gray represents external connections and organizations.

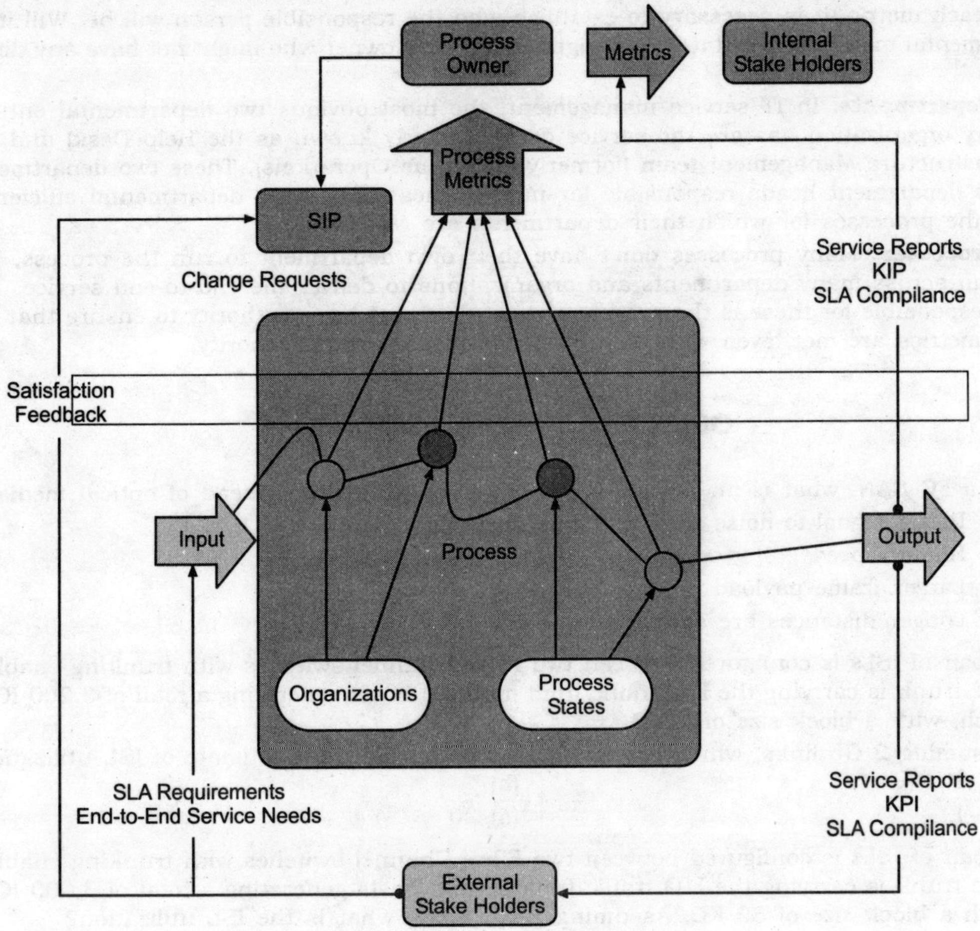

Fig. 7.2. General Process Schematic

Q.7. Why Metrics?

Ans. • Cheshire Puss—she began, rather timidly, as she did not at all know whether it would like the name: however, it only grinned a little wider.

- Come, it's pleased so far—thought Alice, and she went on. Would you tell me, please, which way I ought to go from here?
- That depends a good deal on where you want to get to—said the Cat.
- I don't much care where—said Alice.
- Then it doesn't matter which way you go—said the Cat.
- So long as I get somewhere—Alice added as an explanation.
- Oh, you're sure to do that—said the Cat—if you only walk long enough.

Q.8. Where to use metrics?

Ans. IT service management software tools provide many possible metrics, most of which will be interesting to people in the relevant departments or running the relevant processes. If only these metrics are relied upon then the 'tail wags the dog'. The organization knows what the designer of the software thought it should know, not what business IT alignment demands.

For this reason it is important to start at the other end, design the processes for your organization, then design the metrics to measure them. After this, it is possible to decide how to implement them in software.

For each metric, it is necessary to establish who the responsible person will be. Will it be a departmental manager or will it be a designated process owner who might not have any direct reports.

1. Departments: In IT service management, the most obvious two departmental entities that every organization has are the service desk (formerly known as the Help Desk) and the ICT Infrastructure Management team (formerly known as Operations). These two departments will have department heads responsible for metrics measuring their departmental efficiency, and for the processes for which their departments are responsible.

2. Processes: Many processes don't have their own department to run the process, but rather run across many departments and organizations to deliver the end-to-end service. The person responsible for these is the process owner who must have authority to ensure that the process metrics are met, even with only influence management authority.

OBJECTIVE TYPE QUESTIONS

1. In a FC SAN, what is an advantage of using copper media instead of optical media?
   (a) Better signal-to-noise ratio for short distance
   (b) Higher speed
   (c) Larger frame payload
   (d) Longer distances are supported
2. A pair of ISLs is configured between two Fibre Channel switches with trunking enabled. The trunk is carrying the I/O traffic from multiple hosts generating a total of 3,200 IOPS each, with a block size of 4 KB.
   Assuming 2 Gb links, which percentage represents the best estimate of ISL utilization?
   (a) 1 (b)
   (c) 1 (d) 20
3. A pair of ISLs is configured between two Fibre Channel switches with trunking enabled. The trunk is carrying the I/O traffic from several hosts generating a total of 3,000 IOPS with a block size of 32 KB. Assuming 2 Gb links, what is the ISL utilization?
   (a) 3% (b) 25%
   (c) 50% (d) 90%
4. In a requirements analysis, what are classified as external requirements?
   (a) Availability and reliability (b) Delivery and standard
   (c) Portability and interoperability (d) Privacy and legislative
5. What is a benefit of Kerberos over other authentication protocols?
   (a) Client can provide its own privilege information, reducing authentication traffic
   (b) Ensures a user will not be disconnected from a share because tickets do not expire
   (c) Stores file system ACLs for the domain environment
   (d) User ID and password information is never stored on the Kerberos server

ANSWERS

1. (a) 2. (b) 3. (b) 4. (d) 5. (a)

CHAPTER

8

INFORMATION STORAGE ON CLOUD

8.0. INTRODUCTION

Cloud computing is a computing paradigm, where a large pool of systems are connected in private or public networks, to provide dynamically scalable infrastructure for application, data and file storage. With the advent of this technology, the cost of computation, application hosting, content storage and delivery is reduced significantly.

Cloud computing is a practical approach to experience direct cost benefits and it has the potential to transform a data center from a capital-intensive set up to a variable priced environment.

The idea of cloud computing is based on a very fundamental principal of "reusability of IT capabilities'.

The difference that cloud computing brings compared to traditional concepts of "grid computing", "distributed computing", "utility computing", or "autonomic computing" is to broaden horizons across organizational boundaries.

Forrester defines cloud computing as:

"A pool of abstracted, highly scalable, and managed compute infrastructure capable of hosting end-customer applications and billed by consumption."

Inside this Chapter

Fig. 8.1. Conceptual view of cloud computing

8.1. CLOUD COMPUTING MODELS

Cloud providers offer services that can be grouped into three categories.

1. **Software as a Service (SaaS):** In this model, a complete application is offered to the customer, as a service on demand. A single instance of the service runs on the cloud & multiple end users are serviced. On the customers side, there is no need for upfront investment in servers or software licenses, while for the provider, the costs are lowered, since only a single application needs to be hosted and maintained. Today SaaS is offered by companies such as Google, Salesforce, Microsoft, Zoho, etc.
2. **Platform as a Service (Paas):** Here, a layer of software, or development environment is encapsulated and offered as a service, upon which other higher levels of service can be built. The customer has the freedom to build his own applications, which run on the providers infrastructure. To meet manageability and scalability requirements of the applications, PaaS providers offer a predefined combination of OS and application servers, such as LAMP platform (Linux, Apache, MySql and PHP), restricted J2EE, Ruby etc. Google's App Engine, Force.com, etc. are some of the popular PaaS examples.
3. **Infrastructure as a Service (Iaas):** IaaS provides basic storage and computing capabilities as standardized services over the network. Servers, storage systems, networking equipment, data centre space, etc. are pooled and made available to handle workloads. The customer would typically deploy his own software on the infrastructure. Some common examples are Amazon, GoGrid, 3 Tera, etc.

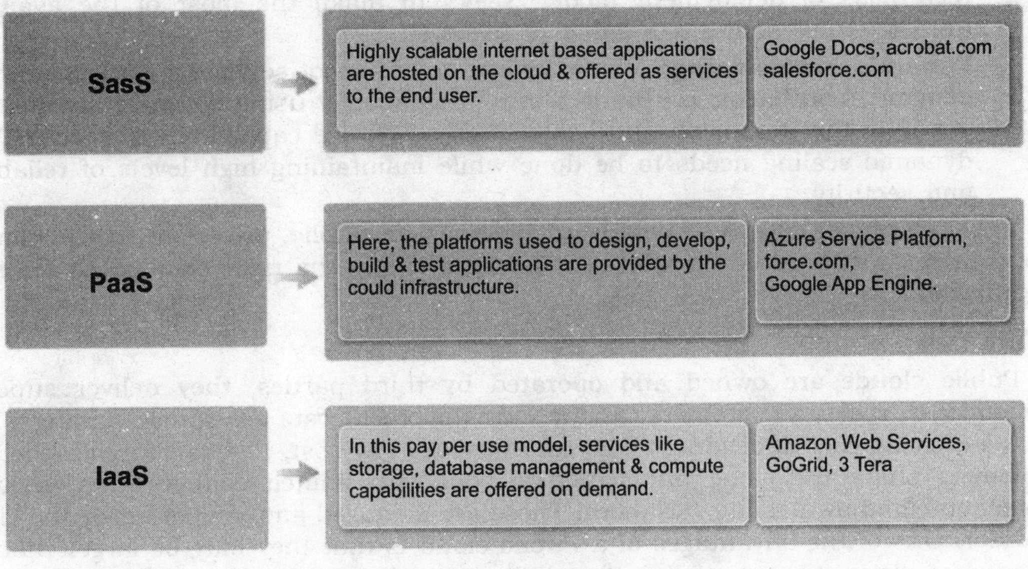

Fig. 8.2. Cloud models

8.2. CLOUD COMPUTING

The term "cloud", as used in this white paper, appears to have its origins in network diagrams that represented the internet, or various parts of it, as schematic clouds. "Cloud computing" was coined for what happens when applications and services are moved into the internet "cloud." Cloud computing is not something that suddenly appeared overnight; in some form it may trace back to a time when computer systems remotely time-shared computing resources and applications. More currently though, cloud computing refers to the many different types of services and applications being delivered in the internet cloud, and the fact that, in many cases, the devices used to access these services and applications do not require any special applications.

Many companies are delivering services from the cloud. Some notable examples as of 2010 include the following:

- **Google:** Has a private cloud that it uses for delivering many different services to its users, including email access, document applications, text translations, maps, web analytics, and much more.
- **Microsoft:** Has Microsoft® Sharepoint online service that allows for content and business intelligence tools to be moved into the cloud, and Microsoft currently makes its office applications available in a cloud.
- **Salesforce.com:** Runs its application set for its customers in a cloud, and its Force.com and Vmforce.com products provide developers with platforms to build customized cloud services. But, what is cloud computing ? The following sections note cloud and cloud computing characteristics, services models, deployment models, benefits, and challenges.
- **Characteristics:** Cloud computing has a variety of characteristics, with the main ones being:
- **Shared Infrastructure:** Uses a virtualized software model, enabling the sharing of physical services, storage, and networking capabilities. The cloud infrastructure,

regardless of deployment model, seeks to make the most of the available infrastructure across a number of users.

- **Dynamic Provisioning:** Allows for the provision of services based on current demand requirements. This is done automatically using software automation, enabling the expansion and contraction of service capability, as needed. This dynamic scaling needs to be done while maintaining high levels of reliability and security.

Enterprises can choose to deploy applications on public, private or hybrid clouds. Cloud Integrators can play a vital part in determining the right cloud path for each organization.

Public Cloud

Public clouds are owned and operated by third parties; they deliver superior economies of scale to customers, as the infrastructure costs are spread among a mix of users, giving each individual client an attractive low-cost, Pay-as-you-go model. All customers share the same infrastructure pool with limited configuration, security protections, and availability variances. These are managed and supported by the cloud provider. One of the advantages of a Public cloud is that they may be larger than an enterprises cloud, thus providing the ability to scale seamlessly, on demand.

Private Cloud

Private clouds are built exclusively for a single enterprise. They aim to address concerns on data security and offer greater control, which is typically lacking in a public cloud. There are two variations to a private cloud:

- *On-premise Private Cloud:* On-premise private clouds, also known as internal clouds are hosted within one's own data center. This model provides a more standardized process and protection, but is limited in aspects of size and scalability. IT departments would also need to incur the capital and operational costs for the physical resources. This is best suited for applications which require complete control and configurability of the infrastructure and security.
- *Externally hosted Private Cloud:* This type of private cloud is hosted externally with a cloud provider, where the provider facilitates an exclusive cloud environment with full guarantee of privacy. This is best suited for enterprises that dont prefer a public cloud due to sharing of physical resources.

Hybrid Cloud

Hybrid Clouds combine both public and private cloud models. With a Hybrid Cloud, service providers can utilize 3rd party Cloud Providers in a full or partial manner thus increasing the flexibility of computing. The Hybrid cloud environment is capable of providing on-demand, externally provisioned scale. The ability to augment a private cloud with the resources of a public cloud can be used to manage any unexpected surges in workload.

8.3. CLOUD COMPUTING BENEFITS

Enterprises would need to align their applications, so as to exploit the architecture models that Cloud Computing offers. Some of the typical benefits are listed below:

1. **Reduced Cost:** There are a number of reasons to attribute Cloud technology with lower costs. The billing model is pay as per usage; the infrastructure is not

purchased thus lowering maintenance. Initial expense and recurring expenses are much lower than traditional computing.

2. **Increased Storage:** With the massive Infrastructure that is offered by Cloud providers today, storage and maintenance of large volumes of data is a reality. Sudden workload spikes are also managed effectively and efficiently, since the cloud can scale dynamically.
3. **Flexibility:** This is an extremely important characteristic. With enterprises having to adapt, even more rapidly, to changing business conditions, speed to deliver is critical. Cloud computing stresses on getting applications to market very quickly, by using the most appropriate building blocks necessary for deployment.

8.4. CLOUD COMPUTING CHALLENGES

Despite its growing influence, concerns regarding cloud computing still remain. In our opinion, the benefits outweigh the drawbacks and the model is worth exploring. Some common challenges are:

1. **Data Protection:** Data Security is a crucial element that warrants scrutiny. Enterprises are reluctant to buy an assurance of business data security from vendors. They fear losing data to competition and the data confidentiality of consumers. In many instances, the actual storage location is not disclosed, adding onto the security concerns of enterprises. In the existing models, firewalls across data centers (owned by enterprises) protect this sensitive information. In the cloud model, Service providers are responsible for maintaining data security and enterprises would have to rely on them.
2. **Data Recovery and Availability:** All business applications have service level agreements that are stringently followed. Operational teams play a key role in management of service level agreements and runtime governance of applications. In production environments, operational teams support:
   - Appropriate clustering and Fail over
   - Data Replication
   - System monitoring (Transactions monitoring, logs monitoring and others)
   - Maintenance (Runtime Governance)
   - Disaster recovery
   - Capacity and performance management

   If, any of the above mentioned services is under-served by a cloud provider, the damage and impact could be severe.
3. **Management Capabilities:** Despite there being multiple cloud providers, the management of platform and infrastructure is still in its infancy. Features like Auto-scaling for example, are a crucial requirement for many enterprises. There is huge potential to improve on the scalability and load balancing features provided today.
4. **Regulatory and Compliance Restrictions:** In some of the European countries, Government regulations do not allow customer's personal information and other sensitive information to be physically located outside the state or country. In order to meet such requirements, cloud providers need to setup a data center or a storage site exclusively within the country to comply with regulations. Having such an infrastructure may not always be feasible and is a big challenge for cloud providers.

With cloud computing, the action moves to the interface that is, to the interface between service suppliers and multiple groups of service consumers. Cloud services will demand expertise in distributed services, procurement, risk assessment and service negotiation areas that many enterprises are only modestly equipped to handle.

8.5. STORAGE ON CLOUD

There are many companies providing free online storage. These storage-as-a-service providers have very similar feature set. These companies provide an easy to use online interface to upload or download files from a user's desktop to the servers on the cloud. Typical usage of these sites is to take a backup of files and data. Other usages are hosting images, videos or large files and making them available from web sites.

All these sites follow a freemium business model. They provide a basic online account that typically has limited disk storage and/or a limit on individual file size. A less restrictive storage scheme is available for paying members. This is a very competitive space with no clear leader. There are as many as 11 sites that provide free online storage, among the top 500 most popular sites. This is as per alexa traffic statistics as on Dec 2009.

Table below gives comparison of free offering of storage-as-a-service providers:

Free Storage as a Service				
Vendor	**File Size**	**Storage Size**	**Waiting time**	**Advertising**
RapidShare	200 MB	Unlimited	Yes	Yes
MediaFire	200 MB	Unlimited	nil	Yes
MegaUpload	2 GB	200 GB	25 sec	Yes
Hotfile	400 MB	Unlimited	30 sec	No
4Shared	500 MB	10 GB	Yes	Yes
DepositFiles	2 GB	Unlimited	60 sec	Yes
YourFileHost	25 MB	?	Yes	Yes
Zshare	1 GB	?	20 sec	Yes
easy-share	200 MB	Unlimited	Yes	Yes
Imgur	10 MB	Unlimited	No	Yes
Uploading	1 GB	Unlimited	Yes	Yes

There are other constraints of the free version. Some of the important ones are explained below.

	Constraint of free storage as a service providers
Vendor name	**Limitations**
Rapidshare	Files not accessed within three months are deleted
MediaFire	No URL is provided to access the files directly from a website
Hotfiles	30 minutes waiting time between two downloads. Files not accessed within 90 days are deleted
4Shared	In the free version, files not accessed for 30 days are deleted

Uploading	Deletes files that are not downloaded after six months in the free version
Imgur	Files more than 1MB are compressed. Files not accessed within three months are deleted
Easy-share	Premium account allows users to download statistics, rename or organize your files into folders or set your file properties
Yourfilehost	Deletes files that are not downloaded in fourteen days
Depositfiles	Deletes files that are not downloaded in thirty days

Table below gives comparison of basic premium version offering of storage-as-a-service providers. Imgur and yourfilehost only have a free version.

Premium version of Storage providers				
Vendor	**File Size**	**Storage Size**	**Waiting time**	**Price**
RapidShare	2 GB	Unlimited	nil	6.99 Euro
MediaFire	2 GB	Unlimited	nil	$6.97
MegaUpload	Unlimited	Unlimited	nil	$9.99
Hotfile	400 MB	Unlimited	nil	$9
4shared	5 GB	100 GB	nil	$9.95
depositfiles	2 GB	Unlimited	nil	$10.95
Zshare	2 GB	?	nil	No
Easy-share	1 GB	Unlimited	Yes	$11.95
Uploading	2 GB	Unlimited	nil	$9.95

An important criteria in selecting a provider for storage space on web is the download speed. However it is difficult for thecloudtutorial to do this comparison, as the download speed is dependent on many factors. Users will need to select couple of vendors based upon information provided above, and then choose one based upon performance.

Comparison of Online Image Storage Providers

There are a large number of companies that provide the ability to upload/share/view images on the web. These organizations provide free online image storage. These sites use freemium business model. The free version of these products typically has some restrictions on size of images. Typical features of online image storage provided by these sites include:

(*a*) Host images and videos
(*b*) Images hosted at these sites may be accessed from blogs and websites
(*c*) Organizing photos into albums, tagging photos
(*d*) Searching images based upon a tag
(*e*) Provide web services API to perform operations on images stored in the web site
(*f*) Setting access control on images
(*g*) Providing comments on images

In this article, thecloudtutorial has reviews of six leading online image storage sites.

1. Flickr (owned by Yahoo)
2. Fotolog

3. Photobucket (owned by Fox interactive media)
4. Webshots (owned by American Greetings)
5. Imagebam
6. Ziddu

A comparison of free version of these sites is included below.

Free Image Storage provider			
Vendor	**Storage Size**	**Video**	**Advertising**
Flickr	100 MB per month; 10MB per image	2 videos	Yes
Fotolog	One image per day	n/a	Yes
PhotoBucket	500 MB; 1MB per image	Less than 5 minutes per video	Yes
Webshots	100 images per month	10 videos per month	Yes
Imagebam	3MB per image	n/a	Yes
Ziddu	200 MB	n/a	Yes

Flickr, owned by yahoo, is the most popular site for storing images online. In addition to restriction on space for monthly upload, user can access only a maximum of 200 photos in the photostream. The disk space restriction of 100 MB per month cannot be increased by deleting images.

Photobucket has monthly bandwidth limit of 10GB on data transfer traffic. Ziddu does not have a premium version. The table below provides a comparison of the premium version of online image storage providers.

Comparison of Premium Image Storage provider			
Vendor	**Storage Size**	**Video**	**Charges**
Flickr	Unlimited; 20 MB per photo	Unlimited no of videos; Limit of 90 seconds max, 500MB per video.	$24.95 annually
Fotolog	Six images per day	n/a	$2.75 for two weeks
Photobucket	Unlimited; 20 MB per image	10 minutes per video and 300MB	$24.95 annually
Webshots	500 images per month	50 videos per month	$19.99 annually
Imagebam	3MB file size limit	n/a	$9.95 annually

Technologies Related to Cloud Computing

Cloud computing typically has characteristics of all these technologies:

(*a*) Grid computing
(*b*) Virtualization
(*c*) Utility computing
(*d*) Autonomic computing

A quick overview of these technologies is given here.

Grid Computing

Grid computing involves a network of computers that are utilized together to gain large supercomputing type computing resources. Using this network of computers large and complex computing operations can be performed. In grid computing these network of computers may be present in different locations.

A famous Grid Computing project is Folding@Home. The project involves utilizing unused computing powers of thousands of computers to perform a complex scientific problem. The goal of the project is "to understand protein folding, misfolding, and related diseases".

Virtualization

Virtualization introduces a layer between Hardware and operating system. During the sixties mainframe started supporting many users using virtual machines. These virtual machines simulated behavior of an operating system for each user. VMWare launched a product called VMware Workstation in 1999 that allows multiple operating systems to run on personal computers.

The virtualization forms the foundation of cloud technology. Using virtualization, users can access servers or storage without knowing specific server or storage details. The virtualization layer will execute user request for computing resources by accessing appropriate resources.

Typically server utilization in data centers can be as low as 10%. Virtualization can help in significantly improving server utilization.

Utility Computing

Utility Computing defines a "pay-per-use" model for using computing services. In utility computing, billing model of computing resources is similar to how utilities like electricity are traditionally billed. When we procure electricity from a vendor, the initial cost required is minimal. Based upon the usage of electricity, electricity companies bills the customer (typically monthly). In utility computing billing is done using a similar protocol.

Various billing models are being explored. A few common ones are:

1. Billing per user count. As an example if an organization of 100 people uses Google's gmail or Microsoft Live as their internal email system with e-mail residing on servers in the cloud, Google/Microsoft may bill the organization on per user basis.
2. Billing per Gigabyte. If an organization is using Amazon to host their data on the cloud, Amazon may bill the organization on the disk space usage.
3. Billing per hour/day. As an example a user may pay for usage of virtual servers by time utilized in hours.

In reality pricing on cloud computing can be very complex. As an example pricing of Amazon S3 as on November 2009 is explained below. Amazon charges for using US S3 are divided into three parts—storage charges, data transfer charges and charges for number of requests. These charges are summed together to compute the total billing.

S3 storage charges		
	US server	**Europe server**
First 50 TB / month	$0.150 per GB	$0.180 per GB
ncxt 50 TB / month	$0.140 per GB	$0.170 per GB
next 400 TB /month	$0.130 per GB	$0.160 per GB
over 500 TB / month	$0.120 per GB	$0.150 per GB

Data transfer charges are further divided into data transfer input and data transfer output. Data transfer rate for incoming data $0.100 per GB. Data transfer rate for outputing data is explained in the table below.

S3 data transfer out charges		
	US server	**Europe server**
First 10 TB / month	$0.170 per GB	$0.170 per GB
next 40 TB / month	$0.130 per GB	$0.130 per GB
next 100 TB /month	$0.110 per GB	$0.110 per GB
over 150 TB / month	$0.100 per GB	$0.100 per GB

S3 charges for requests made		
	US server	**Europe server**
Per 1,000 PUT, COPY, POST, or LIST requests	$0.01	$0.012
Per 10,000 other requests	$0.01	$0.012

Utility computing helps in reducing initial investment. As the computing requirements for an individual or an organization changes, the billing changes accordingly, without incurring any additional cost. If the usage has reduced, then billing will also reduce accordingly.

Autonomic Computing

Autonomic computing is an initiative started by IBM in 2001. Autonomic means "self-managing" computers. In Autonomic computing, computers can automatically correct themselves without human intervention. As an example consider is a network of computers running a set of programs. When there is a hardware failure on one of the computers on the network, the programs running on that computer are "transferred" to other computers in the network. This is an example of "self-correction" or autonomic computing. The analogy typically used is that of human biological systems. Our biological systems take action in self-correcting mode without our explicit knowledge. In the same way the goal of autonomic computing is for computing infrastructure to self-correct itself in unforeseen situations.

8.6. CLOUD VOCABULARY

The benefits ascribed to cloud span both business and technology spheres. For business leaders, your ability to maintain or gain agility and your management options are greatly enhanced. For an IT department trying to deliver services to support the business, cloud services offer a new way to architect and source solutions. By finding cloud services that match noncore delivery of IT services, the IT department can concentrate on finding and delivering the best services for the business more effectively.

At a high level the benefits of cloud services can be categorized as:

- Agility
- Business focus
- Cost and budget control
- Scalability and capacity management
- Governance and compliance

- Security
- Optimized infrastructure
- Isolation
- Mobility
- Refactorization

Agility

From a business perspective, there is much more to consider today beyond your ability to manage your core business and deliver great and timely products and services. Today, competitive pressures, marketing challenges, budget issues, and more are considerable requirements. Your ability to manage situations quickly and efficiently is the key.

The biggest benefit of cloud computing to business today can be framed in terms of agility. Cloud services can offer huge savings in terms of time (for example, when IT capabilities must be delivered quickly). Scaling up or down with cloud services does not usually require additional hardware or software. Cloud services offer minimal setup time, minimal time to scale, and less cash outlay. This is because as a business model, cloud service providers generally host massively scaled systems' capacity that can be switched on upon request.

Suppose, for instance, that you need to scale rapidly for a new project or a seasonal rush. Companies can model these situations using internal resources, but likely at some point they will need to expand beyond that capacity. A decision is made whether to use an external provider to fill the gap; in the world of cloud services, this is called cloud bursting.

Cloud services, as a concept, are available over Internet technologies and enable us to interact or consume them from almost anywhere on any device. While issues of form factor and communication speed create some limitations today, the business benefit of being able to bring key resources to bear on a critical or time-sensitive problem is recognized as a huge benefit to the agility of any business. Having mobile and remote capabilities allows organizations to recruit employees/contractors who can deliver but who cannot or will not travel to their physical locations. Popularized in the 1990s, offshoring was a first example of this business transition: Business services could be offered from anywhere. However, the advent of cloud services means that more capabilities are available to you and to those providers (and from them, too).

Cultural issues aside, web conferencing services, such as the pioneering WebEx, show how product demonstrations no longer require someone to be physically in the room to represent the company. There are now a multitude of meeting options ranging from Citrix's GoTo services for remote access and support, to HP's Halo room for the "in the same room" meeting experience, and to Skype for making video and audio calls worldwide for much less than traditional telecommunication carrier costs. All of these examples illustrate the opportunity to use technology to deliver business results faster and at highly cost-effective price points.

Servicing your customers at scale is possible only through improvements to the scale and functionality of your service and support capabilities. Support over the Internet is one way to do this. In this scenario, you either expose your support model through a web application or you use a provider who will manage that support through a web application on your behalf. Outsourcers such as EDS/HP Enterprise Services, Centerbeam, and more have been offering these types of support services for more than a decade, but the ability to focus support into web-based solutions decreases the number of staff required to answer phones and deal with people directly.

Business Focus

By using the best service from a cloud service provider, a business can potentially focus more energy and talent on optimizing existing revenue streams and aggressively pursuing the development of new ones.

For example, cloud services can enable businesses to gather information, ideas, feedback and so forth from a much wider set of sources (such as customers, partners) than was ever possible by traditional means. This approach is known as crowdsourcing.

Popular crowdsourcing approaches have primarily evolved from the world of Web 2.0 solutions. One business that relies on crowdsourcing is Wikipedia, an online encyclopedia. Wikipedia employs a small organization of less than 50 employees, while utilizing several thousand key volunteers and tens of thousands of other contributors from around the world. While some entries are questionable in terms of veracity, substance or even legality, the overall effort resulted in a much more dynamic and comprehensive set of data than traditional printed encyclopedias could ever match.

Yelp offers the ability to source a set of opinions on a wide range of vendors, from restaurants, to retailers, and more. From this, others can view ratings and comments about those vendors and decide whether to use them. Yelp also shows that these types of solutions can be manipulated (for example, when they gain notoriety, or, when not enough people provide opinions).

Consider a company in crisis. On April 20, 2010 British Petroleum's oil drilling platform, the Deepwater Horizon in the Gulf of Mexico suffered a series of catastrophic failures and collapsed into the water with devastating results to life and nature. The amount of oil escaping was immense. Estimates ranged wildly from 5 to 200 thousand barrels of oil a day, flowing non-stop for over three months. The point here is that BP used crowdsourcing as one approach to deal with the cleanup efforts by creating a "Deepwater Horizon suggestions". As of July over 20,000 suggestions had been submitted, and at least 10 had been tested for use. The US federal government also set up a site with information on volunteering to help with the clean-up effort. This is not to say that crowdsourcing is all perfect. Using services in the cloud like this opens up the potential for anyone in the world with Internet access to "join in" with the crowd. The majority of participants are likely to offer positive input, whereas other individuals or groups are less valuable; some are trolls seeking to make noise, and some are vandals seeking to abuse the system. Mitigation against these and similar issues is centered on access controls, the ability to curate the input, verification processes, and so forth.

Vendors such as Ning, Big Tent, SocialGo, and many others enable for communitybased social networking solutions, even crowdsourcing, in a more controlled environment with stricter access controls to the various parts of their services in the cloud. These additional controls can make crowd-based efforts more compelling to community- based or vertical-focused organizations.

Cost and Budget Control

Although the initial costs of using cloud services may appear less, a better expectation should be that cloud services offer more control over costs or better budget management capabilities. Most cloud services enable you to pay on a monthly, weekly, or per-use basis. Choosing a cost-effective cloud service provider can result in significant savings, but more important is finding a cloud service provider than can accurately report usage patterns to you so that you can confirm the accuracy of your billing based on use.

One noted advantage of using public cloud services is the use of operational expenditure (OpEx) over capital expenditure (CapEx). However, understanding the implications of CapEx and OpEx is critical to effectively managing budgets. The difference between buying a house and renting one is the amount of cash that (usually) goes out the door at one time, and the same concept applies here. IT data centers are generally CapExintensive, because they require initial outlays of cash upfront to build out. Cloud services are generally booked as OpEx because they are consumed through a services agreement over time. Although CapEx can be depreciated over time, essentially allowing costs to be defrayed against profits, the initial drain on cash at hand is often seen as detrimental and to be avoided in the business world. Having flexibility of where to spend OpEx versus CapEx enables an organization to better manage expenditure such that the costs occur over time rather than in one big lump sum, which enables for better cash-flow management.

The concept of Green IT illustrates the significant costs borne by many organizations. The operational costs associated with building and maintaining data centers and with scaling power, cooling, and even basic hardware requirements are significant. Over time business workloads fluctuate and data centers are often left fully powered on, off whether the infrastructure is used. For example, some organizations run financial systems all year, but they hit peak load only once a quarter or once a year during financial reporting periods. There are many ways that organizations can look to benefit from cloud deployments as they try to optimize and approach a Green IT model. For their own private clouds, the goal is to optimize their usage requirements and cycling systems when needed. Organizations use community or public cloud services when needed, essentially pushing their workloads to the cloud when their internal capacity is reached, or cloud-bursting. Although many cloud service providers do not provide utilization data, there is an assumption that cloud service providers have optimized their infrastructure and management tools to maximize usage.

In many cases, organizations no longer have to hire people to manage system updates and backups and therefore can save on staffing costs and on storage management.

Scalability and Capacity Management

The capability to scale quickly to extreme capacity is not a common attribute or goal of traditional data centers and IT departments. As a result, many businesses experience times when they are hampered by a process that takes weeks or months, sometimes even years, to get new compute resources and applications online. This is despite that many large enterprises have already spent millions on computer hardware and have thousands of servers under management (at least on the books). Cloud computing promises the capability to scale massively in terms of systems, connections, bandwidth, storage, and more on an almost immediate basis. The converse is also important, where the service will shrink back down just as quickly if the need is no longer there. These benefits support cost management goals and enable you to grow in line with your requirements in a linear rather than a reactive mode.

The appearance of unlimited capacity is quite appealing. Cloud services need to be tuned to carefully manage actual capacity against expected requirements, yet deal with unexpected demands when necessary, too. Predictive analysis is critical in this respect, and offering those delivery and management capabilities to customers is something that both IT and cloud service providers must do to deliver this benefit to businesses.

Governance and Compliance

Governance is the process used to ensure that regulations, rules, and mandates are followed within an organization. Compliance refers to the facility to monitor and validate that the organization is meeting the governance requirements. Although cloud services introduce potential challenges in matching these requirements, the opportunities to improve an organization's governance and compliance stance can be greatly improved.

Taking a service-based approach to delivering both business and IT functionality that incorporates the governance processes allows for closer alignment to compliance reporting. Sourcing specific, metered services from cloud service providers that specifically match governance models allows for better reporting, more accurate alignment between service usage and fluctuations in business workflow, and ultimately, faster time to market. This also brings up the need for a service catalog and related management tools to ensure usage matches expectations. We will talk about this in chapters 7 through 10 as this closely relates to overall operations management and security as well.

In some cases, organizations have found that by using a third party to manage their IT services they are better able to deliver services within their governance models and compliance mandates. For small businesses in particular, the cost of hiring to manage these requirements can be significant itself, much like security. Using pooled resources and services allows business to capitalize on the best of that common capability.

It is important to note that although cloud services provide potential improvements to your governance and compliance position, you cannot abdicate responsibility for compliance. While different models of cloud computing architecture actually vary the amount of "control" you and the service provider have over the service, none of them changes your responsibilities to protect data, privacy, or service levels.

This is a critical point made that is reiterated throughout the text: Although you can look to the cloud for specific capabilities and functions to support governance, compliance, and even security, you cannot abdicate these obligations. For example, if a breach of your customer or employee data occurs, the responsibility to report and rectify remains yours. The cloud service provider may help deal with the issue. The cloud service provider may suffer some of the same financial repercussions. However, the cloud service provider is not the owner of the problem. Therefore, any reliance on the cloud's controls and safeguards, and on the governance and compliance practices of a cloud service provider, must be carefully evaluated in line with your governance, and compliance needs to ensure that the solution enhances your posture, instead of increasing either direct or ancillary risks.

Security

Whereas many potential users of cloud services have a visceral or gut reaction that suggest cloud services are not secure, that is certainly not always the case. We discuss security opportunities and challenges in much more detail later in the book, but for now consider many organizations find significant improvements in their security posture by using cloud services. Often, because of their size or financial limitations, small and even medium-sized businesses cannot hire security specialists or respond quickly to security incidents. Using specific best-of-breed capabilities in the cloud to prevent or identify fast-moving or specific threats is a benefit to all organizations, perhaps best illustrated by the vendors who provide antivirus and antispam solutions outside the traditional IT environment.

Optimal Infrastructure

The ability to host multiple capabilities in the same hardware, software, or service allows for the follow-on benefit of optimized use of the infrastructure. Multitenancy as a model for using compute resources has existed since the 1960s, when IBM challenged traditional time-sharing models by adding virtualization with its VM/370 series.

Multitenancy can occur at any or all levels of the architecture, as follows:

- **Virtual layer:** Virtualization provides the ability to create specific environments for each process, application, or operating system. This model isolates everything above the virtual layer itself but allows for the use of pooled resources below that layer, most commonly hardware such as networking, processor, memory, input/ output, and storage resources.
- **Application layer:** The user interfaces of most web applications allow for specific fixed graphical and behavioral elements alongside customized elements associated with a specific organization, individual, or function. If one component or functional element fails in the delivery of the interface, it is easier to replace because those elements derived from other parts of the application and can be easily reconstructed.
- **Database layer:** Data for multiple applications, through to multiple customers, can be stored in the same database, and thus allow for the focus on optimal data structures rather than entire infrastructures to support each individual requirement.

To gain the best advantage of cloud requires that each of these architectural layers be considered carefully to determine the best layer or layers to optimize for multitenancy.

Isolation

Almost a corollary to multitenancy is both the ability and requirement for isolation at each of the layers. While an application vendor such as Salesforce.com may use a common database layer for managing customer data, it is essential that the security mechanisms around the application layer maintain isolation between the various customers. Fundamental for any type of cloud provider is the requirement to offer isolation at each level of service exposure to their cloud users. This may be the infrastructure, platform, or software applications. In addition, APIs and management tools must also ensure that isolation exists in terms of identity management and access models, key management and encryption, and user interfaces. This is a specific set of technical requirements that need to be carefully evaluated when using a third-party solution in a private cloud and in any public cloud.

Mobility

The idea of web-based services has been around for many years, as have application hosting and outsourcing. The ability to get to the services from anywhere from any device has been a goal that is finally being broadly realized, and it threatens business models and IT departments who must contend with the security issues associated with data being available on devices either temporarily or long term.

In August 2009, ABI Research released a report that said mobile cloud computing subscribers would total nearly one billion by 2014. The ABI report contained the following reasoning:

There are two primary reasons why ABI believes cloud computing will become a disruptive force in the mobile world. The first is simply the number of users the technology has the power to reach: far more than the number of smartphone users alone. The second reason has to do with how applications are distributed today. Currently, mobile applications are tied to a carrier. If you want an iPhone app, for example, you have to first have a relationship with the mobile operator who carries the iPhone. If you want a Blackberry app, the same rule applies. But with mobile clouding computing applications, as long as you have access to the web, you have access to the mobile application.

Although there are many arguments against this position, the general direction in cloud services is to support more open standards and therefore the dependence on specific carriers is certainly less than in years previous. This means that offering employees, customers, and partners better access and links to your organization via an exploding mobile world is absolutely possible.

Refactorization

Some like to think of cloud computing as an opportunity to do away with all their existing infrastructure challenges and costs-a "burn it to the ground" or rebuild scenario. This is truly not the real option, especially for any enterprise with more than a year under its belt with existing IT, and especially not for medium to large enterprises with many current or legacy systems in place. The reality is that cloud services allows an IT department to refactor some or all of their existing systems over time and usually take advantage of cost-effective new ways to deliver IT services to the business as a result.

The same concept can also be applied to data center design. Historically, IT has designed data centers using the model of high availability, focusing on repair as a core requirement instead of considering the best mode for recovery to availability. This includes using certified hardware with comprehensive support and maintenance contracts. This includes concepts such as fail-fast, highly integrated systems, and deploying on stable QA tested solutions on a prepared basis with massive change control to ensure the ability to roll back failures.

Large cloud service providers are approaching infrastructure design with a qualitatively different approach. They focus on low-cost commodity hardware where possible. Faults should be simply routed around until some standard recovery can be achieved on a scheduled or even ad hoc basis. This is much more cost efficient in many modes, yet requires a change in thinking that may benefit IT. Modeling can be important here, but if you are not prepared to consider alternative approaches, you will never get to model it. Also note that it is often difficult to model third-party cloud services well, given the potentially vast array of failures that can occur. Traditional IT approaches do not immediately match well to these new architectures and will require some refactoring. We all know that hardware will fail, networks will fail, and an entire data center may fail. Truly, any part of the stack, including the human components (from operators to users), may fail in some way. So what does that mean? We need to monitor different things. It is common to model small failures in a larger system and monitor for those failures (e.g., disk out of space, router offline). Cloud services are modeled as services, and therefore if you refactor how to manage them in this context, you begin to see dependencies across the whole system rather than in terms of just the data center itself, which in turn allows you to focus on core services rather than all services. This change can truly impact IT's ability to deliver and maintain high-availability services.

Summary of Benefits

Using cloud services to supplement or replace IT functions should allow an IT department to deliver more innovative capabilities to the business by focusing more on the service delivery and less on the hardware and software updates. Most benefits derive from the pooled nature of cloud services being offered through multitenancy architectures (or more simply, economies of scale). Costs, risks, controls, and more are aggregated across thousands of customers rather than one individual organization's data center.

Many definitions of cloud computing identify self-service or self-provisioning as a benefit. Although self-service can help make cloud services easier to use, the reality is that self-service requires a mode of operations that includes service catalogs, automated provisioning and de-provisioning, and more to be effective. Therefore, we consider selfservice as something that although beneficial in small environment is in reality a result of delivering good service-oriented architectures that may or may not be cloud based in nature.

In summary, cloud computing provides significant opportunities. Thinking these through in relation to your own business challenges is important. Perhaps even more critical, you want to remember that your competitors are also considering how to capitalize on these opportunities.

However, cloud computing has challenges. Your organization may already be ready to adopt and adapt to new technologies or ways of achieving your business goals, but finding success with cloud services may require a mindset that allows you to change how you manage risk and control. To achieve success with cloud, regardless of internal or external options, IT needs to be the service broker and aggregator for the business, providing guidance, cost management, and governance in this new model.

8.7. APPLICATIONS AND SERVICES ON CLOUD

The service models of cloud computing as defined by NIST are shown in Fig. 8.3.

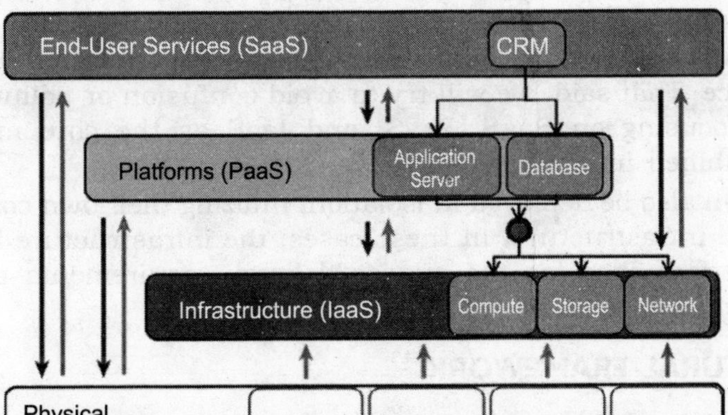

Fig. 8.3. Cloud service models based on the NIST definition

Each of the service models offers different levels of capabilities and responsibilities to the provider and the consumer of the service. In addition, these service models may utilize a custom architecture, or the physical infrastructure, to exist. These services may rely on a service offering from a lower service model. In this way, cloud services in the higher levels, those being platform and software, may be entirely built atop other cloud services. The most compelling way to think about these models is that if everything has a set of APIs from the lowest to the highest levels, the service-orientated nature of cloud services becomes very clear, as follows:

- **Software as a service (SaaS):** The capability provided to the consumer is to use the provider's applications running on a cloud infrastructure. The applications are accessible from various client devices through a thin client interface such as a web browser (e.g., web-based e-mail). The consumer does not manage or control the underlying cloud infrastructure, including network, servers, operating systems, storage, or even individual application capabilities, with the possible exception of limited user-specific application configuration settings.
- **Platform as a service (PaaS):** The capability provided to the consumer is to deploy onto the cloud infrastructure consumer-created or acquired applications created using programming languages and tools supported by the provider. The consumer does not manage or control the underlying cloud infrastructure, including network, servers, operating systems, or storage, but has control over the deployed applications and possibly application hosting environment configurations.
- **Infrastructure as a service (IaaS):** The capability provided to the consumer is to provision processing, storage, networks, and other fundamental computing resources where the consumer is able to deploy and run arbitrary software, which can include operating systems and applications. The consumer does not manage or control the underlying cloud infrastructure but has control over operating systems, storage, deployed applications, and possibly limited control of select networking components (e.g., host firewalls).

To elaborate the potential use of lower services by higher level services, Fig. 8.3 includes an example SaaS customer relationship management (CRM) application that resides entirely atop the various cloud services deployment options, using an application server and database in a PaaS scenario, which in turn rely on an IaaS option for compute, network, and storage capabilities.

Critically, in Fig. 8.3, we also introduce the concept of XaaS. XaaS represents anything as a service. That said, we will try to avoid confusion or pollution of the "aaS" nomenclature by focusing on SaaS, PaaS, and IaaS as the core models for cloud services, often combined into its own acronym of SPI.

These models can also be deployed in isolation, utilizing their own compute, network, storage, and related infrastructure. In these cases, the infrastructure is architected to deliver the best performance for the service delivery requirements and not generic capabilities as defined by each service layer.

8.8. ARCHITECTURAL FRAMEWORK

The deployment models of cloud computing as defined by NIST are shown in Fig. 8.4 and are described in the following list.

- **Private cloud:** The cloud infrastructure is operated solely for an organization. It may be managed by the organization or a third party and may exist on premises or off premises.
- **Community cloud:** The cloud infrastructure is shared by several organizations and supports a specific community that has shared concerns (e.g., mission,

security requirements, policy, and compliance considerations). It may be managed by the organizations or a third party and may exist on premises or off premises.

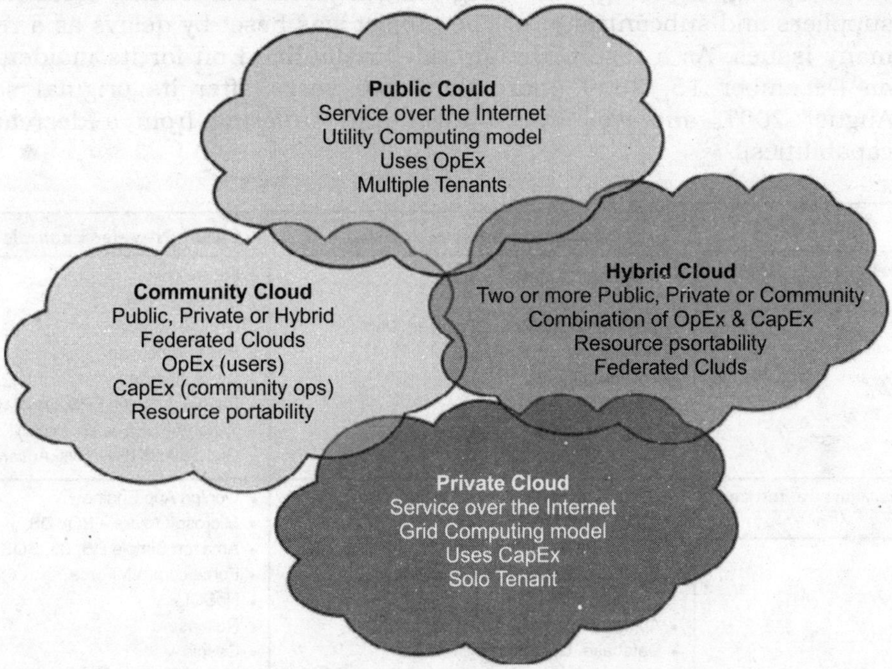

Fig. 8.4. Cloud deployment models based on the NIST definition

- **Public cloud:** The cloud infrastructure is made available to the general public or a large industry group and is owned by an organization selling cloud services.
- **Hybrid cloud:** The cloud infrastructure is a composition of two or more clouds (private, community, or public) that remain unique entities but are bound together by standardized or proprietary technology that enables data and application portability (e.g., cloud bursting for load balancing between clouds).

Cloud architecture sample capabilities and industry examples:

- Most of the examples are public cloud options, and the opportunity to fashion your own solution such as these as private cloud options does negate some of the opportunities we have discussed so far.
- The addition of cloud-driven processes and web-based services to the SPI framework is intended to illustrate the higher-level business models and processes that can be considered. Furthermore, here we illustrate the term mashup, which comes from the Web 2.0 environments. A mashup is a composite web-based service created by mixing together other web-based services, or more appropriately, higher-level cloud services. This creates useful and varied solutions for users, but also introduces management challenges as delivery is affected via distributed capabilities. This essentially becomes a chained service model, and as with all integrations, or should we say "mashups," things can go wrong across the service components.
- For a business example more grounded (ahem) in supply chain management, consider the history of the Boeing 787 Dreamliner. Planned to be the most

advanced technical and component-based airplane in large-scale commercial aviation, Boeing approached the effort by taking a nontraditional approach to collaborating, sourcing, and integrating myriad components from hundreds of suppliers and subcontractors. The project was beset by delays as a result of too many issues. As a result, the aircraft finally lifted off for its maiden test flight on December 15, 2009 (more than two years after its original schedule of August 2007, and well over budget and suffering from a decreased set of capabilities).

Cloud Service	Service Attributes/Description	Cloud Provider Examples
Software as a Service (SaaS)	• Consumer Websites • Mashups • Multi-Tenant Business Focused Web Application • Collaboration • CRM, SFA, ERP etc • Management Applications/Intercaces • APIs for Service Integrations	• Flicker.com • Myspace.com • Google Apps • SelesForce.com • Cisco WebEd • Paypal, Amazon FPS, DevPay • Yahoo APIs (Search, Flickr) • Google Apls (Payment, Adsense)
Platform as a Service (PaaS)	• Application Environments – J2EE, RoR, .Net • Mashups of SaaS Services • Configuration Platform and Scripting Engines • Hosted Application Development Environments • Application Infrastructure Capabilities • Database, Data Stores • Message Queues • APIs for PaaS Integrations	• Goolge App Engine • Microsoft Azure – SQL DS • Amazon Simple DB, S3, SQS • Force.com, VMForce • NoSQL. • Rollbase • Caspio • Hadoop
Infrastructure as a Service (IaasS)	• Virtual Servers • Compute Capacity Enabled by API Based • Provisioning • Logical Disks • Networking • APIs for IaaS Service Integrations	• Amzon EC2 • Elastra • Rackspace • Savvis • FlexScale • RightScale • GoGrid • CloudSwitch • SNIA

Fig. 8.5.

- History shows that there were numerous problems for Boeing in terms of supply chain management: integration due to standardization issues, component shortages, qualitycontrol issues, and more. Similar issues come to the forefront when an organization chooses to use cloud services without clear strategic goals and management tools being in place.
- *To succeed with cloud services, business and IT leaders must recognize and deal with the fact that the role of IT is changing to include much more comprehensive supply chain management and vendor management. If cloud services are to be used, the traditional IT team makeup is incomplete and so must be enhanced with more legal, contractual, and business expertise. Those familiar with outsourcing are in a much better place to manage this new approach to delivering business value.*

- *Cloud services can enable the business to gain much greater control over its ITdependent decisions, as long as it has the correct management processes and tools in place. This is discussed extensively in Part III, "Life in the Cloud-Planning and Managing the Cloud," of this book, where we examine the approaches and requirements for managing cloud services solutions.*

The success of cloud computing is largely based on the effective implementation of its architecture. In cloud computing, architecture is not just based on how the application will work with the intended users. Cloud computing requires an intricate interaction with the hardware which is very essential to ensure uptime of the application.

These two components (hardware and application) have to work together seamlessly or else cloud computing will not be possible. If the application fails, the hardware will not be able to push the data and implement certain processes.

On the other hand, hardware failure will mean stoppage of operations. For that reason, precaution has to be done so that these components will be working as expected and necessary fixes has to be implemented immediately for prevention as well as quick resolution.

Data Centers

One of the most distinguishing characteristics of cloud computing architecture is its close dependency on the hardware components. An online application is just a simple application that could be launched in different servers but when the application is considered with cloud computing, it will require massive data centers that will ensure the processes are done as expected and timely.

Data centers for cloud computing architecture are not your run-of-the-mill data processing centers. It's composed of different servers with optimal storage capacity and processing speed. They work together to ensure that the application will be operating as expected. The area is usually in a highly controlled environment where it would be constantly monitored through various applications and manually checked for actual physical problems.

The data center could be considered as the backbone of cloud computing architecture. The destruction of it could easily mean millions of dollars in additional spending for companies. For that reason, data centers of large companies are often kept secret to avoid infiltration either by hacking or actual physical damage.

8.9. APPLICATIONS IN CLOUD COMPUTING ARCHITECTURE

Enabling the capacity of the data centers is the software that does the processing. With the help of the data centers, the processing time will be fast as the speed of transaction will be suggested by the hardware capabilities of the data center.

The application in cloud computing will call on the assistance of the hardware not only in processing but also in data gathering. Although it would be possible that data will come from another source, data centers will usually house the data in their server farms for faster access and easier processing.

The challenge for applications in cloud computing is largely based on the number of requests the application could handle. Although this factor could be highly suggested by the data center, the application will usually have a threshold if they are not properly written.

To deal with this concern, developers use metadata to enable personalized services to their users as well as data processing. Through metadata, individualized requests will be entertained and will be properly implemented. Metadata also ensures uptime of transaction as data requests will be slowed down if the developer chooses to do so.

The Unix Philosophy

Cloud computing architecture and its implementation is strongly defined by the Unix Philosophy. It's basically a set of rules and principles that developers would have to follow to ensure that cloud computing would be easily implemented and the application would be for the advantage of the user.

Although there are many definitions, rules and principles for Unix Philosophy, they all point out to one belief: build one thing and make sure it works consistently.

Through the Unix Philosophy, developers who design the architecture for cloud computing have to remember that they are there only to support only one application or at least an output. Cloud computing could be composed of different stages but these stages are geared towards the consistent computing of the online application.

Keeping up with the demands of the application is the data centers and server farms. It could be said that too much hardware could be used to support a single process but these are implemented to ensure that the application would have as many back-up plans as it could have.

Applications from cloud computing is also implemented with the thought of consistent security and performance monitoring. Through proper implementation of cloud computing architecture, the application will be accessible 24/7 with a new 100% uptime for their users.

8.10. APPLICATIONS AND SERVICES ON CLOUD

Successful implementation of cloud computing requires proper implementation of certain components. Without any of these components, cloud computing will not be possible. These components can't be easily implemented by one person alone.

Cloud computing will require persons with different expertise, experiences and backgrounds. As it will require more people in the industry, it's no wonder why cloud computing is a very expensive venture. But even with the expenses that the company would often have to spend, the advantages provided by cloud computing is far more than the initial spending.

Some would resort to a cloud computing vendor because of the lack of resources while others have the resources to build their cloud computing applications, platforms and hardware. But either way, components have to be implemented with the expectation of optimal performance.

8.10.1. The Client–The End User

Everything ends with the client. The hardware components, the application and everything else developed for cloud computing will be used in the client. Without the client, nothing will be possible.

The client could come in two forms: The hardware component or the combination of software and hardware components. Although it's a common conception that cloud computing solely relies on the cloud (internet), there are certain systems that requires pre-installed applications to ensure smooth transition. The hardware on the other hand will be the platform where everything has to be launched.

Optimization is based on two fronts: the local hardware capacity and the software security. Through optimized hardware with security, the application will launch seamlessly.

8.10.2. The Service–the Functions in Cloud Computing

Cloud computing always has a purpose. One of the main reasons cloud computing become popular is due to the adoption of businesses as the easier way to implement business processes. Cloud computing is all about processes and the services launched through cloud computing always has to deal with processes with an expected output.

The optimization on services is based on two things: The proper development of the application and the end user. Sometimes, the service could be used by the user wherein their experience is greatly affected by their gadget.

Cloud computing has been the byword for different businesses today. This is a type of process that relies o the internet or online connectivity for different data processing instead of using the local gadget. The internet has come a long way of providing different types of services to users.

Years ago, the internet is just used merely for information gathering and email. Today, the internet could become a host of different types of applications which will not require any local installation in the users end.

Cloud computing promises portability as users would just have to need a strong internet connection to ensure the process is done. Cloud computing could even emulate the desktop experience.

Cloud computing is usually focused on the enterprise. As businesses today require more collaboration in real time without any geographical consideration, cloud computing became a viable option as this will provide real time interaction in business process. It offers the portability to different users without having to constantly require the local gadget for certain installations.

8.10.3. Considering Web 2.0

The term "Web 2.0" on the other hand, have been used for many years long before cloud computing became popular in the software development industry. But instead of defining Web 2.0 about the components it requires, Web 2.0 is more about the interaction the user will receive.

In Web 2.0, the focus of different online applications is the user wherein they are given the freedom to connect and make some changes on their online environment. Web 2.0 is more about the interaction between users.

Classic examples of Web 2.0 are the different online social networking websites. Users of these websites will be able to connect to one another while trying to make some changes in their online environment depending on their preferences.

Although Web 2.0 will still require hardware support, its focus is more on the actual interaction of the online application with the user. The application seen in different Web 2.0 websites are always geared towards the user's need to do something about their experience online.

8.10.4. Cloud Computing Behavior

The behavior of cloud computing is highly dynamic wherein the only way the process would be possible is through proper interaction of the application and hardware. If one of the components in cloud computing will not work or at least will execute below par, cloud computing will never work. Developers and business managers have to make sure everything is according to plan which will never falter in any occasion. Certain support measures have to be implemented to prevent any form of downtime. Infrastructure and extensive monitoring is usually a requirement to properly implement cloud computing and stay true to its behavior.

8.10.5. From Desktop to Browser

One of the best characteristics of cloud computing is its ability to remove the need of desktop or local applications. Everything could be extracted online and everything could be performed in a secured browser. Major browsers have the capability to handle different types of cloud computing and they will work seamlessly as long as the supporting factors could properly execute.

Obviously, cloud computing is all about doing everything online. This means cloud computing will require different factors should be working together. Aside from these factors, they should be working consistently to ensure continuous business operation.

8.10.6. Ability to be Dynamic

Getting the application, data and proper process in the cloud is not just getting the data available in the server and could be extracted by request. The processing power of data center and application should have dynamism that should be observed at all times. Dynamism in cloud computing is the ability to redistribute the processing power of the cloud at will. If there are few users who tries to use the cloud, the resources of the cloud should be distributed on those number of users. At the same time, cloud computing should be redistributed when the number of users will increase.

8.10.7. Distribution of Resources

Aside from the ability to adapt to the number of users and data requests, cloud computing should have the ability to work with different form of resources. Most well known service providers do not rely their operations in one service center alone. They would usually come with two or more server farms - infrastructure with multiple and massive servers.

These are not done just to store large amount of data, multiple server farms in different areas are redistributed in locations so that they could provide ample support to other data centers. If one server farm will go down, other server farms will take over the load temporarily.

8.10.8. Operations as Abstraction

Cloud computing is never a simple concept. Each application that will run in the server and would eventually become online will require different set-up. This is essential in abstraction which will enable the application to be highly dynamic as well as adaptable. Abstraction should be observed as behavior in cloud computing as this will provide the needed operations for cloud computing. Through abstraction, the application should be adaptable enough to different scenarios. This is useful especially for businesses that requests adaptable processes.

8.10.9. Platform Development

The only way the application in cloud computing should be launched is through a platform. By using a platform such as browser, the local gadget will only require minimal application to be launched. One classic example is the browser wherein online applications could be easily launched.

This is very challenging for developers since launching the application online will require greater adaptability. Browser incompatibility is still one of the biggest challenges a developer must face. Internet Explorer, Firefox, Opera, Safari and other major browsers have different implementation to different functions. This means developers might not have the ability to implement certain application in a specific browser.

8.10.10. Physical Requirements

Getting everything done online requires powerful hardware capabilities. As already indicated, the online application in cloud computing requires different type of data center. Everything will be for nothing if the application will not have powerful support.

The physical requirement is actually the most challenging part in cloud computing. The physical requirement will need considerable amount of spending just to make sure everything will work seamlessly. Once everything has been set-up, additional support for software as well as hardware components should be there. The security of these components is always a requirement for disaster prevention and optimum performance.

Implementing cloud computing through a platform is one of the most popular options for businesses today in online transactions. Largely different from software based cloud computing, platforms are basically programming languages or applications that could be customized based on the need of the enterprise. Because the platforms are geared towards different functionalities needed by the enterprise, they are dubbed as platform as a service or PaaS.

Before going further, it is important to differentiate PaaS to SaaS (Software as a Service). SaaS are applications that could be used in the could by different enterprise. They already have predefined functions and the enterprise would only need to adapt to these functions. Paas on the other hand, provides the basic platform wherein developers and the enterprise have to design from the scratch or the preloaded functions.

8.11. ESSENTIAL CHARACTERISTICS OF CLOUD COMPUTING

Full Application Development Cycle: PaaS is not just launching an application online. It will require planning, coding and testing before they are fully implemented for proper use. This will take time and resources which should be expected since PaaS is keen on implementing customized functions and services for the specific enterprise.

Use of Online Programming Language: PaaS is an online application. Naturally, it will require programming languages made for online interaction. From simple HTML to highly complicated JavaScript, and Java, applications could be used to build online applications which will serve cloud computing.

Powerful Integration: The online application built by developers should never be the final version. Updates should be available and different forms of integration should be available for the application. This is necessary for PaaS as the application will require the application to be highly intricate with data hosted in the server. Through integration, Mash-up of different application is possible.

Collaboration and Instrumentation: The development of PaaS should not be limited to a team of developers alone. Adaptation of the application is very important since it will ensure ease of development for other developers as well as maintenance when the new developer takes over.

Through collaboration, instrumentation of functions becomes possible. This will even give developers and PaaS providers to sell certain functions.

The forms PaaS could be easily distinguished based on how they are developed and the requirements a vendor would provide to the enterprise.

Cloud computing is about moving services, computation, and data-for cost and business advantage-to an internal or external, location-transparent service or services. By making data, services, capacity, and more available in a service-based model, they can be much more easily incorporated into business and IT processes that can be ubiquitously accessed. This is often at much lower cost, increasing the value to the business by enabling opportunities for enhanced collaboration, integration, and analysis on a shared common platform.

From just these few definitions, we see both commonality and disparity, and those same issues exist across all the many and varied approaches out there. As a result, there is a multitude of overlapping definitions of cloud. The analysts have them, the marketers have them, users have them, and standards groups have them, and so on. We have one, too, but we won't get wrapped up in debates about it. Instead, we focus on the capabilities that the cloud model provides.

Perhaps most aggravating when investigating cloud services options is the marketing approach to relabel or brand anything and everything "as a service." This is called cloud washing. This just becomes confusing initially and an exercise in futility in the long run as terms become overloaded. We see "storage as a service" and "security as a service" competing acronym-wise with "software as a service." So, one of the advances is to say that each of these options fits into one of the three delivery models, SPI, and be done with it.

- The cloud services market is developing at a rate faster than most other technology and business models.
- Cloud services are changing business models and industries and are creating new opportunities for all.
- There are many definitions of cloud services, and the NIST definition is the most widely accepted definition used today.
- The benefits of cloud services span both business and technology spheres.
- Cloud is not a panacea.

8.11.1. Development through Preloaded Functions

There are platform as a service providers that concentrate in providing developers preloaded functions. They are often considered as proprietary functions wherein the work of developers is only limited to getting the functions together. The vendor will provide everything and the developers will just have to study the functions and how they could be related to the need of the enterprise.

8.11.2. Development through Web Hosting

This type of service in PaaS is the most basic in this form of cloud computing. Developers will have to look for a programming language that could be launched in the server, build different functions and integrate the data located in the server. The role of PaaS provider is to host the functions and prevent it from possible downtime. This type of service concentrates in ensuring the data is available 24/7 and the functions will work as expected.

8.11.3. Development through Frameworks

There are developers who opted to use frameworks in building an application. These frameworks are tools in building application wherein the coding is on the native programming language of the developer. But with the help of the framework, the native coding will be translated to another programming language fit for PaaS. Good examples of this type of development are frameworks for Ajax based applications and Ruby on Rails.

8.11.4. Advantage/Disadvantages of Cloud Computing Platform

The main advantage of platform development through cloud computing is on the cost of development and deployment. Since the applications that come from PaaS will be launched online, little to no requirement will be made on the client side.

On the other hand, cloud computing platform might not have the ability to adapt to different changes on the demand of the enterprise. This is especially true when the application is made through proprietary function from the vendor. Developers have to carefully inspect the history of updates of the vendor to ensure compatibility to the different needs of the enterprise.

8.12. CLOUD SERVICE PROVIDERS AND MODELS

Many companies are delivering services from the cloud. Some notable examples as of 2010 include the following:

- **Google:** Has a private cloud that it uses for delivering many different services to its users, including email access, document applications, text translations, maps, web analytics, and much more.
- **Microsoft:** Has Microsoft Sharepoint online service that allows for content and business intellligence tools to be moved into the cloud, and Microsoft currently makes its office applications available in a cloud.
- **Salesforce.com:** Runs in application set for its customers in a cloud, and its Force.com and Vmforce.com products provide developers with platforms to build customized cloud services. But, what is cloud computing? The following sections node cloud and cloud computting characteristics, services models, devployment models, benefits, and challenges.

Characteristics

Cloud computing has a variety of characteristics, with the main ones being:

- **Shared Infrastructure:** Uses a virtualized software model, enabling the sharing of physical services, storage, and networking capabilities. The cloud infrastructure, regardless, of deployment model, seeks to make the most of the available infrastructure across a number of users.
- **Dynamic Provisioning:** Allows for the provision of services based on current demand requirements. This is done automatically using software automation, enabling the expansion and contraction of service capability, as needed. This dynamic scaling needs to be done while maintaining high levels of reliability and security.

8.13. CLOUD SECURITY AND INTEGRATION

Security is one of the biggest concerns of businesses in any form. Whether a business is a small brick-and-mortar or a multi-million online ventures, security should be implemented. Exposing the company to different security flaws is always inviting to different elements with malicious intent. A single security strike could mean millions of dollars for businesses and might single handedly close the business down.

Proper implementation of security measures is highly recommended for cloud computing. The mere fact that the application is launched through internet makes it vulnerable to any time of attack. An application available in LAN (Local Area Network) only could even be infiltrated from the outside so placing an application over the internet is always a security risk. This is the unique situation of cloud computing. Implementation of cloud computing could require millions of dollars in infrastructure and applications development but it still places itself at risk for different types of attacks.

Protecting the Users

Above everything else, cloud computing or any type of online application format should consider protecting its users. Developers should make sure that data related to the user should not be mishandled and could be extracted just by one.

There are two ways to ensure cloud computing security: restrictive user access and certifications.

Restrictive access could come from simple username/password challenge to complicated CAPTCHA log in forms. But applications in cloud computing should not only base itself on these challenges. IP specific applications and user time-outs are only some of the security measures that should be implemented.

The challenge in restrictive user access is to limit the access privilege of the user. Each user will have to be assigned manually with security clearance to ensure limitation of access to different files.

Certifications are also important for user certification. Developers have to open their application to security specialists or companies that provide certifications for security. This is one way of assuring users that the application has been fully tested against different types of attacks. This is often the dilemma for cloud computing as external security checks might open the company secrets on cloud computing. But this has to be sacrificed to ensure the security of their users.

Data Security

Aside from user protection against different types of attacks, the data itself should be protected. In this aspect, the hardware and software linked to cloud computing should be scrutinized. Again, a certification is highly desired in this part of cloud computing.

The hardware component for cloud computing on the other hand requires a different type of security consideration. The location of data center should not only be selected because of its proximity to controllers and intended users but also on its security (and even secrecy) from external problems. The data center should be protected against different types of weather conditions, fire and even physical attacks that might destroy the center physically.

With regards to the hardware component in relation to the application, certain manual components have to be available for increased security. Among them is manual shutdown to prevent further access of the information. Although data could be controlled with another application that data could be infiltrated unless the application is shutdown immediately.

Recovery and Investigation

Cloud computing security should not only focus itself on prevention. Ample resources should also be focused on recovery if the unfortunate event really strikes. Even before disaster happens, certain plans have to be in place to ensure that everyone will be working in unison towards recovery. The plans do not have to be focused on software attacks alone–certain external disasters such as weather conditions should have separate recovery plans.

When everything has been recovered, developers and the company handling the application should have the means to investigate the cause of the problem. Through investigation, certain conditions that lead to the event could be realized and insecurities could be discovered. Even legal actions could be done if security has been breached on purpose.

Security is one of the most difficult task to implement in cloud computing. It requires constant vigilance against different forms of attacks not only in the application side but also in the hardware components. Attacks with catastrophic effects only needs one security flaw so it's always a challenge for everyone involved to make things secured.

SUMMARY

1. **Software as a Service (SaaS):** In this model, a complete application is offered to the customer, as a service on demand. A single instance of the service runs on the cloud & multiple end users are serviced. On the customers side, there is no need for upfront investment in servers or software licenses, while for the provider, the costs are lowered, since only a single application needs to be hosted & maintained. Today SaaS is offered by companies such as Google, Salesforce, Microsoft, Zoho, etc.
2. **Platform as a Service (Paas):** Here, a layer of software, or development environment is encapsulated & offered as a service, upon which other higher levels of service can be built. The customer has the freedom to build his own applications, which run on the provider? Infrastructure. To meet manageability and scalability requirements of the applications, Paas providers offer a predefined combination of OS and application servers, such as LAMP platform (Linux, Apache, MySql and PHP), restricted J2EE, Ruby, etc. Google?s App Engine, Force.com, etc are some of the popular PaaS examples.
3. **Infrastructure as a Service (Iaas):** IaaS provides basic storage and computing capabilities as standardized services over the network. Servers, storage systems, networking equipment, data centre space, etc. are pooled and made available to handle workloads. The customer would typically deploy his own software on the infrastructure. Some common examples are Amazon, GoGrid, 3 Tera, etc.
4. Cloud computing is a computing paradigm, where a large pool of systems are connected in private or public networks, to provide dynamically scalable infrastructure for application, data and file storage. With the advent of this technology, the cost of computation, application hosting, content storage and delivery is reduced significantly.
5. Cloud computing is a practical approach to experience direct cost benefits and it has the potential to transform a data center from a capital-intensive set up to a variable priced environment.
6. The idea of cloud computing is based on a very fundamental principal of "reusability of IT capabilities". The difference that cloud computing brings compared to traditional concepts of "grid computing", "distributed computing", "utility computing", or "autonomic computing" is to broaden horizons across organizational boundaries.
7. Public clouds are owned and operated by third parties; they deliver superior economies

of scale to customers, as the infrastructure costs are spread among a mix of users, giving each individual client an attractive low-cost, Pay-as-you-go model. All customers share the same infrastructure pool with limited configuration, security protections, and availability variances. These are managed and supported by the cloud provider. One of the advantages of a Public cloud is that they may be larger than an enterprises cloud, thus providing the ability to scale seamlessly, on demand.

8. Hybrid Clouds combine both public and private cloud models. With a Hybrid Cloud, service providers can utilize 3rd party Cloud Providers in a full or partial manner thus increasing the flexibility of computing. The Hybrid cloud environment is capable of providing on-demand, externally provisioned scale. The ability to augment a private cloud with the resources of a public cloud can be used to manage any unexpected surges in workload.
9. Data Security is a crucial element that warrants scrutiny. Enterprises are reluctant to buy an assurance of business data security from vendors. They fear losing data to competition and the data confidentiality of consumers.
10. In many instances, the actual storage location is not disclosed, adding onto the security concerns of enterprises. In the existing models, firewalls across data centers (owned by enterprises) protect this sensitive information. In the cloud model, Service providers are responsible for maintaining data security and enterprises would have to rely on them.

QUESTION WITH ANSWER

Q.1. Just what is Cloud Computing?

Ans. A cloud computing is a general which is used to describe the migration of computer tasks from a traditional infrastructure composed of clients and servers into a model which embraces computing as a utility. In the utility model, assets (computing time, network bandwidth, and even applications) are used and consumed in a manner similar to how electricity or water is consumed and ued you pay for what you use, nor more and no less. Benefits of this model include:

- Businesses can focus financial resources and employees on their business, and not on building and managing computing infrastructure and applications;
- Computing power, applications and network bandwidth are elastic-they come and go dynamically as required and only cost money when used.
- Often, for small and mid-sized organizations, software applications which otherwise would be unaffordable become quite affordable.

Q.2. How is Cloud Computing different?

Ans. A tradditionally, most companies have maintained their own data centers, applications, servers and clients. With this model, companies had high fixed cost for the equipment, software, and the data center itself.

Even worse, it takes many specialized personnel to manage the data center, the networks, and the "stacks" that are associated with each enterprise applications. Cloud computing is a way to cut costs by sharing expensive resources between multiple parties. At the extreme, Cloud computing allows data centers, networks, servers, middleware, software and people to be shared among multiple companies.

Q.3. What are the different cloud models?

Ans. There are three different models that are often called "Cloud Computing":

- Infrastructure as a service (IaaS)
- Platform as a service (PaaS)
- Software as a service (SaaS)

There are variations to each of these but generally any cloud computing model will fall into one of these three categories. We will explain each of these in depth later in this FAQ.

Q.4. What are some trends that are driving people to think about Cloud Computing?

Ans. In an ideal world, many things,

First, the general trend is that computing is migrating towards "web-centric" applications. The client platform and its associated operating system are becoming less important (witness Microsoft fear of Google) as applications become "portable", less dependent or any specific platform (e.g. Windows) and migrate towards "The Web".

Second migrating towards Cloud Computing results in lower CAPEX and OPEX since hardware and software acquisition costs are lowered, and sicne fewer personnel are required to operate "Cloud sourced" applications.

Third ongoing maintenance costs are lower since there is no hardware to maintain, no software to keep updated, and no data center to pay for. Application run in a shared data center, so high data center CAPEX and OPEX is amortized among hundreds or thousands of customers, depending on the specific applications.

Fourth some cloud computing models, especially SaaS, result in true client side platform independence any standard web browser will allow you to run these applications.

Q.5. Can we have a concrete example of Cloud Computing?

Ans. Sure a simple example is afforded by Yahoo mail, which almost everybody is familiar with, Yahoo effectively uses the Software as a Service (SaaS) model, With Yahoo mail (and many other examples), you "plug into" the service just like a utility, with Yahoo mail:

- You do not need servers and on-site storage
- You do not need support staff-its basically self-configurable by the end-user with a minimum of instruction.
- You do not need a client-side program
- You do not need to buy software

  All you need is an internet connection and a web browser. To summarize,with Cloud Computing may offer the following benefits when correctly applied:

- Lower initial CAPEX
- Reduced support staff requirements
- More scalable (both up and down) as demands change
- Ubiquitous access

Q.6. Why is Cloud Computing so efficient?

Ans. Cloud Computing save money for the following reasons:

- It is based on "multi-tenancy" architecture
- One application is shared by multiple users, which allows development and acquisition costs to be distributed across hundreds or thousands of clients.
- Data center costs are shared by many users
- Paid for in predictable monthly installments.

Q.7. What is a stack?

Ans. We will use the term 'stack' to help support our definition of Cloud Computing so we need to explain clearly what it is.

Systems are implemented as a "stack" from bottom to top. Traditionally, companies have owned and implemented this stack. This means large investments in data centers, network, servers, operating systems, support software such as Oracle, Application software, and presentation software.

One of the big problems with this model is that this entire investment must be paid for whether its used to capacity or not. And, when capacity is reached, more capacity must be added at great expense, and, what happens when you only need extra capacity for a short while?

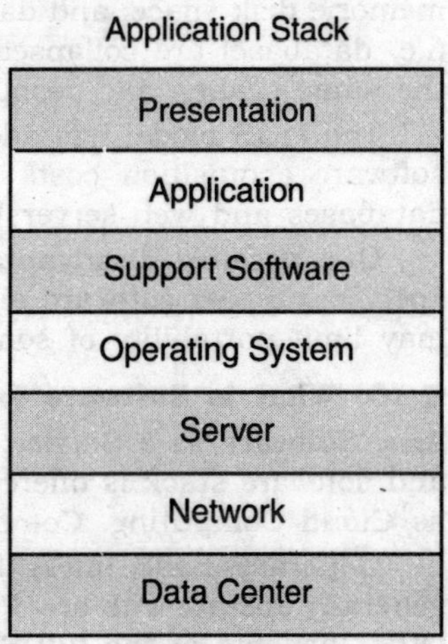

Fig. 8.6.

Cloud computing gives an organization the chance to share these expenses with others. We will see the different models in the next question.

Q.8. What is Infrastructure as a Service (laaS)?

Ans. Infrastructure as a service, or IaaS, is a utility computing model which provides basic infrastructure but no software. These are two subcategories of thie model:

- The traditional "co-location" model which provides the data center facilities and network connectivity.
- The true laaS model, which provides the data center, network, and the computer server platform itself, either as a true system or as a virtual machine.

As can be seen in the illustration to the left, the data center, network and (possibly) the server itself are offered as "Infra-structue".

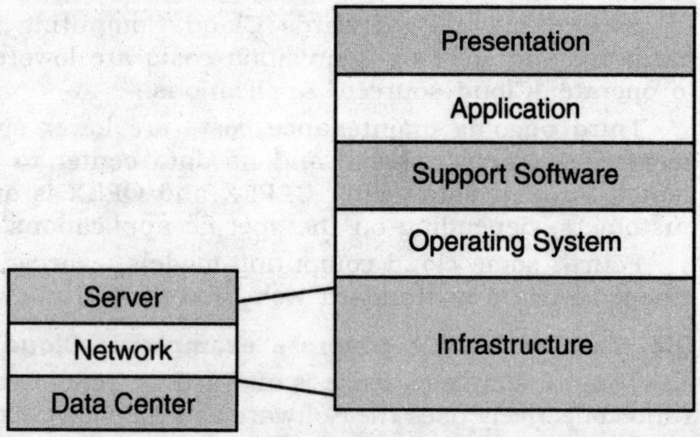

Fig. 8.7.

The true laaS model is especially useful in cases where extra computing resources are needed to satisfy a "burst" in demand. A customer can rent a server from the laaS company and transfer an existing application to the rented servers to meet the burst. When the demand is gone, the rented server can simply be released. Thus, the extra resource is only paid for when needed.

Compare this against traditional approaches. In the traditional 'do it yourself" approach, the extra server and supporting infrastructure would need to be bought and maintained even when not needed-quite an expensive proposition.

To summarize, laaS provides:

- On-demand creation of server and internet capacity.
- Simple creation and configuration of physical or virtual machines.
- Low per-hour system rental and low bandwidth charges.
- Is 'elastic'; scales to demand-need more, buy more need less and it just goes away!
- Is flexible - can run any application.

Q.9. What is Platform as a Service (PaaS)?

Ans. Platform as a service, or PaaS, is utility computing model where the CPU, network memory, disk space, and data center infrastructure, operating system and support applications (i.e. database) are collapsed into an abstraction known as the "Platform". This model enjoys the same scaling and resource sharing advantages as LaaS model.

The PaaS model has additional benefits over the laaS model in that it may allow for lower software acquisition costs for the client becuase expensive support software licences (e.g., databases and web servers) are included as part of the offering.

One possible disadvantage of the PaaS model is that applications must comply with the specific support software and operating system that is provided by the service provider. This may limit portability of some existing applications.

Q.10. What is Software as a Service (SaaS)?

Ans. Software as a Service, or SaaS, is a utility computing model where the entire hardware and software stack is offered as a package. This model is generally what most people think of as Cloud Computing. Commercial examples include Sales Force.com and Google Apps.

Generally SaaS offers lower entry and operational costs over the other models. It will generally operate with any PC that has a standard browser, simplifying client-side configuration. Other benefits of the full SaaS model include:

- Simplifies client maintenance and reduces application level security vulnerability exposure.
- Application acquisition software maintenance, server and data center costs are lowered or justs disappear.
- Collaboration between members of a team, even when that team crosses company boundaries, is improved since all documents and data are maintained in a single shard place.
- Per-seat fee
- Scales up or down as necessary.
- Applications are professionally maintained by SaaS vendor.
- No local applications to patch-all done by the "Cloud".
- Data is backed up by SaaS vendor.
- Some SaaS vendors and applications allow customization, for example for a company to add its own branding.

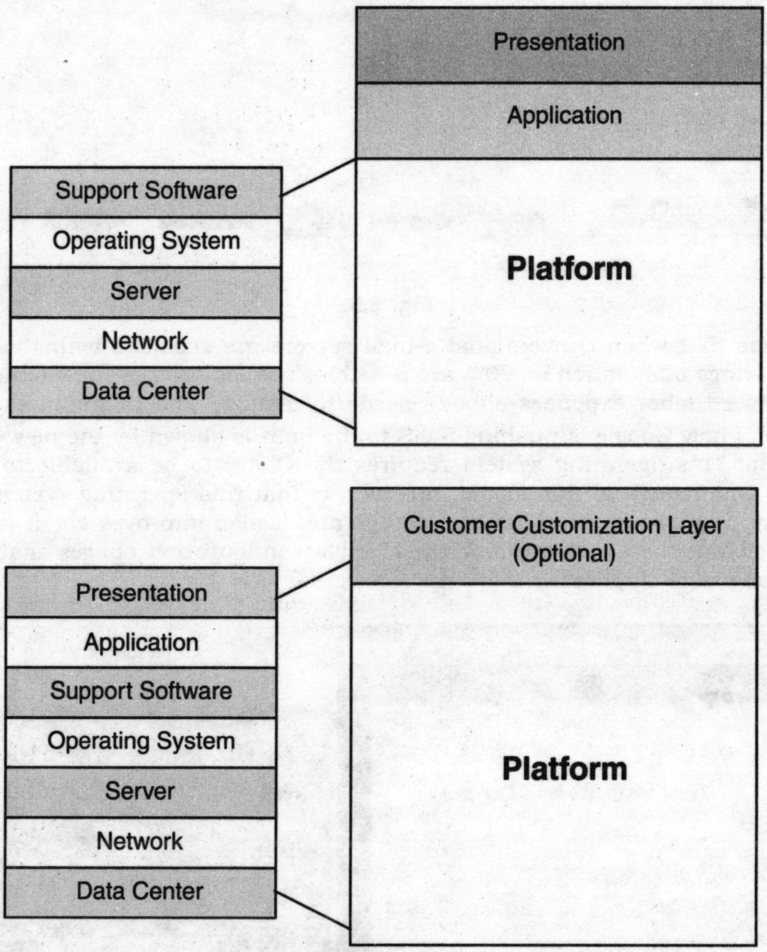

Fig. 8.8

Google claims two million businesses, large and small, are running Google Apps and that this number is growing by thousands daily. One benefit is that when Google apps is used, ones documents are readily available from any client machine that has Internet connectivity. The charge per user is $50 per year, with no long-term contractual requirements.

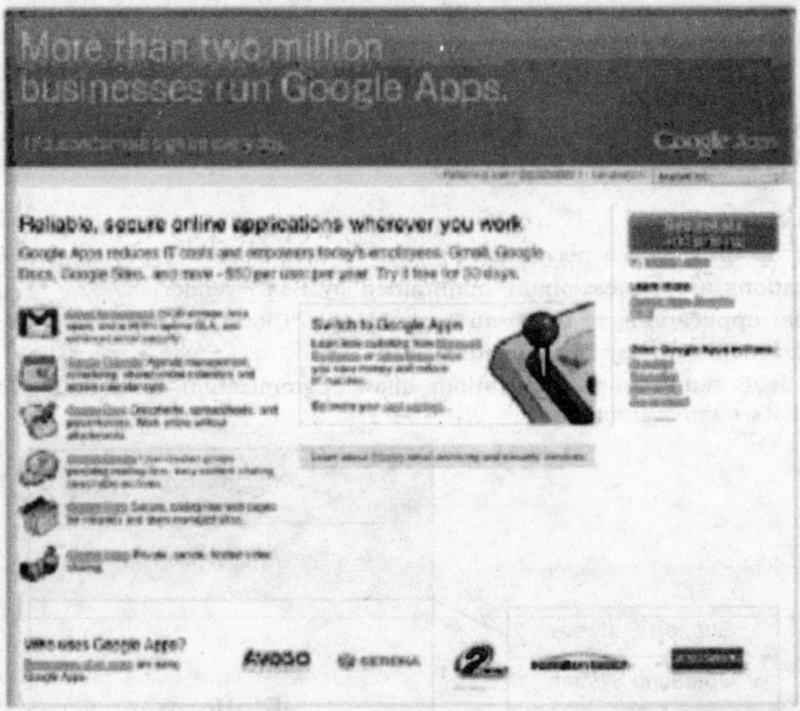

Fig. 8.9.

Google claims that when conventional e-mail servers are replaced with their Cloud based solution, cost savings of as much as 70% are possible. The majority of the claimed cost savings comes from reduced labor expenses-nobody needs to manage and maintain the email server.

An example of how Google is pushing SaaS to the limit is shown by the new Google Chrome operating system. This operating system requires the Cloud to be available to do just about anything at all. One benefit of this model, however, is that this operating system requires very modest (and low cost) client side hardware to operate. it also improves client side security by virtue of the fact that there are very few user visible configuration chages that can be made, and installing any user applications is impossible.

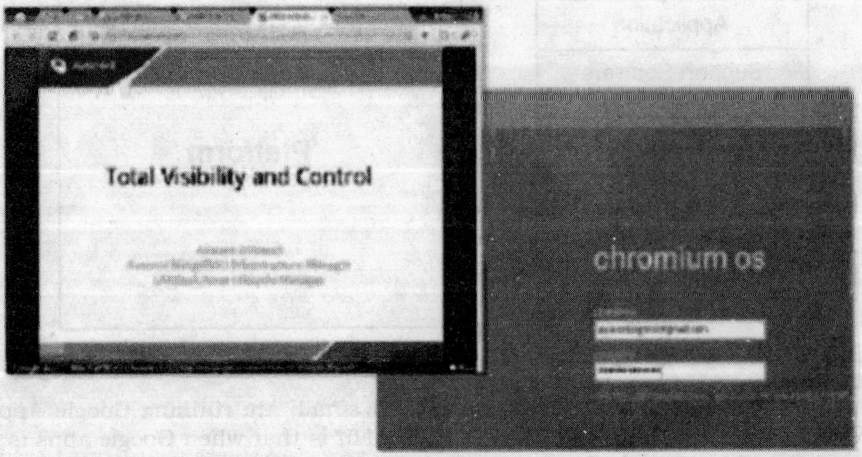

Fig. 8.10.

Shown is an example of the Chrome login screen and the Chrome presentation application. In essence, the client is transformed into a "dumb terminal", completely reliant on the Cloud to operate.

Q.11. When should l use one Cloud model over another?

Ans. Each of the Cloud models have their own benefits which help define where they are best used.

When a specific software stack is required, for example when migrating a legacy application, laaS is the right fit. This model allows capacity to be quickly added for existing applications because it just provides the basic infrastructure on which to run those applications.

When developing, migrating or expanding capacity for web based applications which use common commodity platforms such as Linux, Apache, mysol and Perl(LAMP), PaaS is often the best choice. On the other hand, migrating existing applications which use other platforms may prove to be a more daunting task. An excellent example of an application of PaaS is a Cloud based web server.

SaaS is used when the offered application generally meets the clients needs. For example, SaaS applications which provide common 'office' type applications such as word processors, calandars, Customer Resource Management (CRM) or email are readily available and customizable to meet your companies specific needs.

Q.12. Just what is my company responsible for with the different Cloud models?

Ans. The diagram below helps to explain the different models. The layers of the models are shown and the levels of responsibility or what is owned are shown by the bars to on the right side of the chart.

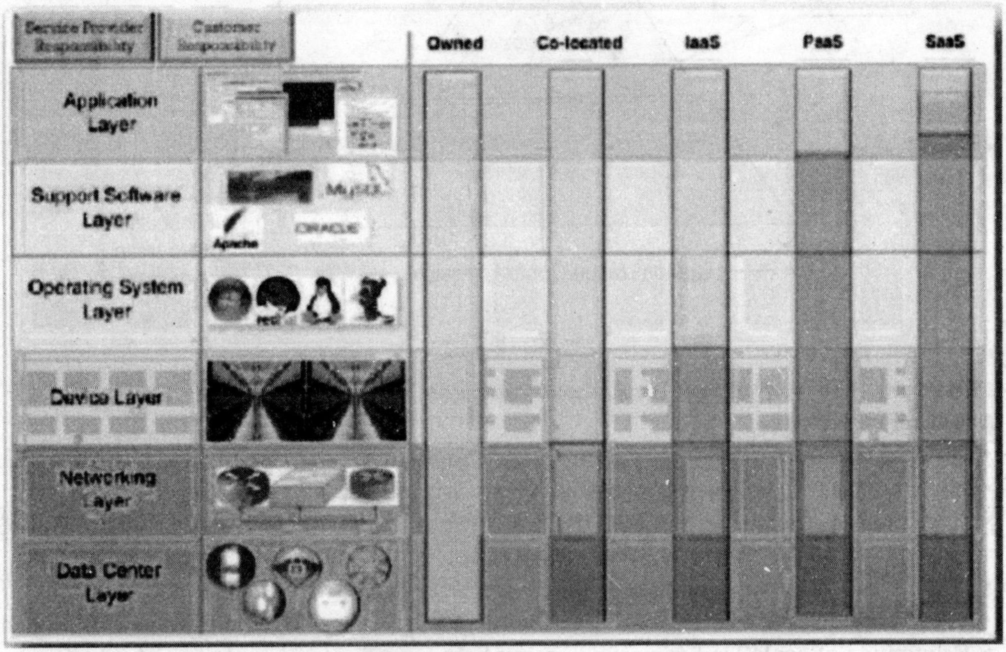

Fig. 8.11.

With the "owned" or traditional model, the customer is responsible for the entire stack from top to bottom. As more functionality is added, the Service Provider becomes responsible for some of the layers of the stack. In the co-located model, the Service Provider is responsible for the data center and the networking layers. in the laaS model, the Service Provider is also responsible for the "device", which could be the physical server hardware or could be a virtual machine.

PaaS provides for the stack, up to the support software layer, while SaaS provides for the entire stack. We show a little overlap in the SaaS model, because insome cases the customer can embellish the supplied application to meet specific requirements.

One really good example of SaaS is provided by Google apps, although this is by no means the only example. Google Apps is an office productivity suite which includes applications such as a word processor, spreadsheet, calendar, collaboration tools, a presentation program, an email application and so-on. All applications are hosted in the Cloud, and all data (files) reside in the Cloud on Google's servers.

Q.13. How much flexibility do the various models of Cloud Computing offer?

Ans. The graphic below helps to illustrate how much flexibility the various models offer, Generally, the higher up the Service Provider chain of value you go, the less flexibility you have. At the bottom of the pyramid, with the laaS model, you have more flexibility as you have the ability to make your own platform choices. With the PaaS model, the platform is chosen for you, but you still have flexibility in terms of the specifics of the application. With the SaaS model, the applications itself is built by the service provider. You may have some customization options but these options will be limited.

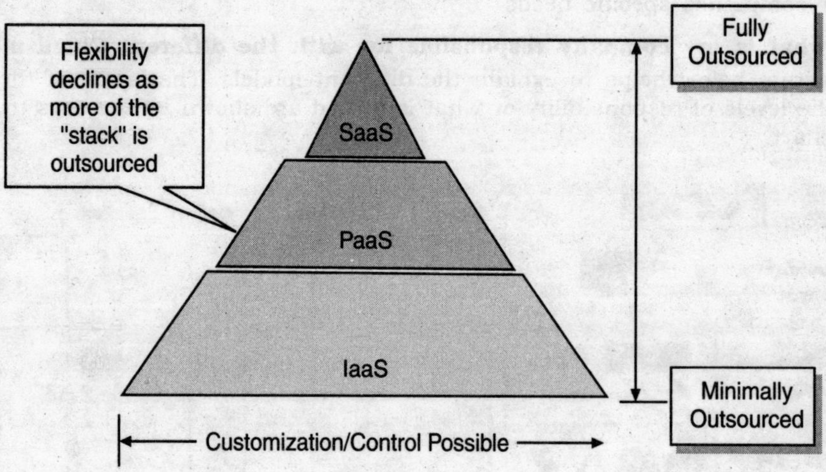

Fig. 8.12.

Q.14. What are the advantages and disadvantages of laaS?

Ans. Advantages include:

- Quick addition of capacity- 'clastic' to grow and shrink as capacity requirements change.
- Most hardware, network and data center infrastructure costs are eliminated.
- Lower labor costs are network infrastructure costs, data center costs are ongoing facility maintenance costs are omitted.
- High availability internet connections from hosting provider.
- Allows existing applications to be moved into the Cloud with a minimum of modification.
- Economies of scale lead towards hosting being resident in word class data centers with tops tier equipment at an affordable price.

Disdvantages include:

- Perceived loss of control of assets and physical security (no longer physically visible, for example).
- Network outages (local connection to internet or the Internet itself) means complete sevice loss.

- More internet bandwidth required from facilities.
- Possible loss of logical security, concerns of information leakage from one customer to another.
- Compliance auditing (i.e., SBOX) may become more complex as there is no direct control of assets located within Cloud.

Q.15. What are the advantages and disadvantages of PaaS?

Ans. Advantages include:

- Quick addition of capacity- 'elastic' to grow and shrink as capacity requirements change.
- Hardware, network and data center infrastructure costs are eliminated
- Some software costgs (OS and Support Software) are eliminated.
- Initial pre-configuration of core-OS and some support software is eliminated
- High availability internet connections from hosting provider
- Allows some existing (platform complaint) applications to be moved into the Cloud with minimum modification.
- Economies of scale lead towards hosting being resident in world class data centers with top tier equipment.

Disdvantage include:

- Less application flexibility as applications must conform to pre-defined platform template.
- Only certain types of applications (i.e. LAMP) are suited for this model, depending on specifies of platform.
- Less portability of existing applications as constrained to a single pre-defined platform configuration.
- Same security/auditing/network bandwidth and dependency concerns as with laaS.
- Provider "lock-in"; portability of applications between providers may prove difficult.

Q.16. What are the advantages and disadvantages of SaaS?

Ans. Advantages include:

- Feature rich applications developed to world class standards and constantly refined.
- Easy and quick set up, often by non-technical people and on demand.
- Easy and quick tear down, on demand when need for the application subsides.
- No long-term contracts means can be used fors short-term bursting.
- Low per-user cost
- Minimal to no development expense; applications are generally, ready to use
- Entire stack acquired and managed by third-party, reducing or eliminating acquistion costs.
- Makes complex applications affordable to companies (e.g. SMes) who would otherwise not be able to afford them
- Applicaitons available from anyplace with internet conneciton
- Low client requirements.
- Flexible client options–operating system and platform agnostic, usually runs from handheld to desktop.

Disdvantages include:

- Often "One Size fits All" customization/flexibility are limited.
- Dependent on connectivity always being available; working offline is difficult or impossible.
- May run slower when Internet is heavily loaded.
- Provider "lock in" is very likely; difficult or impossible to switch vendors.
- Security is a concern; data leakage for example.
- Ability to audit and traceability are concerns.

- Business problems of the cloud provider will become your problem.
- Cloud vendors technical problems will become your problem.
- Security and compliance audits become problematic.

Q.17. If all data center operations are moved to the cloud, what happens to my internet bandwidth requirement?

Ans. Your Internet bandwidth requirement will certainly become an issue. Bandwidth into the Cloud may become a bottleneck, depending on the size of your internet connection, the distance from the cloud service provider, the time of day, and the specific types of applications that are remotely hosted. If the applications that you put into the Cloud are not bandwidth intensive, then things will probably work fine. If they are bandwidth intensive, then you may have to investigate getting dedicated bandwidth allocated between your site and the Cloud service provider. This takes time to provision and certainly costs money.

Q.18. If 1 move all my data into the Cloud, just how secure is it?

Ans. Most cloud vendors will assure you that they have great security and that they are externally audited. Although this is most likely true, for many organizations, it begs another question; are cloud vendors audits transitive to your organization? This is a very important question to get an answer to since for many companies audits of procedures and compliance with government regulations is part and parcel to their business.

Another question is how much much can you really trust a third party? Keep in mind that your companies most secrets lie deep within the bowels of servers and disk drives within a data center that you control. When you push your data into the cloud, your precious bihts are actually sharing the same infrastructure, routers, wires, disk, drives, computers and memory with other companies. These other companies could be your worst enemies. So you are putting a lot of trust into the cloud provider to make sure that your data is isolated from everybody else.

So, whether to trust or not ends up being a serious judgement call for IT staff and senior management. This may motivate keeping the most sensitive data safely within the intranet, and only placing less sensitive data and and applications out to the cloud.

Q.19. Do 1 loose control when I move my datta into the cloud?

Ans. How much control is lost depends on how "deep" you go into the cloud. The higher up the 'stack' you go, the less control and flexibility you have.

You also need to worry about "cloud lock-in", some cloud vendors offering PaaS require custom virtual machines which are NOT transportable to another cloud vendor. This means that if you need to change cloud vendors for some reason, the change could be expensive. It also means that you need to be sensitive to how well your cloud vendor is doing financially.

SaaS is even more susceptible to lock-in. This means you loose your development and training investments if you need to "move".

Q.20. What is a "private cloud"?

Ans. We have held off on discussing private clouds. Now that we explained the different cloud models, we can explain what a private cloud is and how it works.

The concept of a private cloud is simple-you take the concepts of a 'public cloud" and adapt them to live within the intranet. This means that the company has control over the cloud rather a third party-so many of the perceived risks of cloud computing go away, but so do some of the benefits.

According to Gartner Research:

- Public cloud computing is defined as a style of computing in which scalable and elastic IT-enabled capabilities are delivered as a service to external customers using internet technologies.
- Private cloud computing is defined as a style of computing in which scalable and elastic IT-enabled capabilities are delivered as a service to internal customers using internet technologies.

One interesting thing to consider-private Clouds are an excellent way to safely try Cloud Computing. Even better, at some point in the future as the model matures, it may be possible.

Q.21. If public Cloud Computing is so great, why would I need to consider a private Cloud?

Ans. It has been shown that public Cloud Computing has benefits and risks, so public Cloud computing is not the perfect model. It's application needs careful thought. The public Cloud model is perfect for some applications but may need further maturation time for other more risky or sensitive applications. The private Cloud model gives you the ability to enjoy some of the benefits of Cloud Computing while keeping the associated risks safely at bay.

Q.22. What should I think of when I build a private cloud?

Ans. Here are some things to think about if you are considering building your own private cloud:

First, leverage commodity hardware and virtualization technologies. This makes it easier to adapt to the "plug-in" model that is the basis of most cloud models. Commodity hardware is inexpensive, simple and readily available. By embracing virtualization, you allow for the decoupling of the applications from the underlying hardware, and make for simple transportation of applications from one hardware device to another, remember even if the hardware is different, the virtual machine at the application side wil be the same hardware platform when that option is possible.

Make sure that applications are not coupled to any specific hardware implementation. Again, this makes the application portables so that it can easily be migrated into the Cloud.

Leverage Cloud management technologies that allow you to pool computing resources and allocate virtual machines to hardware dynamically. This allows for simple tansport of applications between servers, and between Clouds. It also allow you to quickly adapt to surges in computing demand by simplys allocating more horespower from your server farm to the problem.

You need to think differently. The IT organizaiton needs to think of the data center not as dedicated servers allocated to specific functions, but rather as a matrix of server resources that can be allocated to power a matrix of applications.

Finally, be careful about what applications you migrate to the Cloud model. Some legacy applications simply wont fit, especially those with built-in assumptions about the platform that they run on. What makes sense to keep as, just leave alone. What cannot be put into a public cloud, leave private. And, what can safely be pushed into the public Cloud with minimal risk, push into the public cloud.

Q.23. What do I need to think about relating to managing a private Cloud?

Ans. The management solution needs a way to quickly and efficiently allow you to allocate application workloads to available infrastructure dynamically, and to monitor those servers at a server (physical processor) level such that you can know when the server is working as hard as it efficiently can, Conceptually, you would need to:

"Put 3 instances of the CRM application on Server 43, right now".

You need a way to monitor application performance so that the Cloud can be 'tuned' dynamically as loading grows or shrinks: Conceptually, you would need to:

'Server 43 is working too hard-move an instance of the CRM application onto Server 67 which is not working at all'.

To maximize energy consumption in the Cloud, you need a way to control physical hardware assets so that they are turned on and of as needed and don't waste energy. You need to be able to efficiently monitor and manage Cloud assets from any place.

OBJECTIVE TYPE QUESTIONS

1. What is a remote replication solution for EMC CLARiiON storage systems?

(a) EMC MirrorView/A (b) EMC SnapView
(c) EMC SRDF/AR (d) EMC TimeFinder

2. Which management task eliminates single points of failure in a data center?
(a) Availability (b) Capacity
(c) Performance (d) Security

3. Which local replication technology does EMC TimeFinder/Mirror implement?
(a) Copy on First Access (b) Copy on First Write
(c) Full Volume Mirroring (d) Pointer based

4. Why would you implement an iSCSI solution?
(a) All storage arrays supply GigE ports
(b) Block storage over IP is much faster than FC
(c) iSCSI HBAs are inexpensive
(d) Leverages existing IP network infrastructure

5. What is the name of the drive sub-assembly containing the platters and read/write heads?
(a) ATA (b) SATA
(c) HDA (d) IDE

6. What is the correct order of operations for array-based synchronous replication?
(a) Write is received by the source array from host/server
(b) Remote array sends acknowledgement to the source array
(c) Source array signals write complete to host/server
(d) Write is transmitted by source array to the remote array

7. Which failure does RAID technology guard against?
(a) Disk failure (b) Host Bus Adapter failures
(c) Host failures (d) Switch failure

8. Which statement describes content address (CA)?
(a) A set of functions that enable communication between applications
(b) An identifier that uniquely addresses the content of a file and its location
(c) An identifier that uniquely addresses the content of a file and not its location
(d) An identifier that uniquely addresses the content, quality, condition and other characteristics of data

9. What is a server capacity management task?
(a) Establishing user quotas (b) Installing multi-path software
(c) Maintaining user logins (d) Monitoring faults and events

10. Which statement describes the iSCSI protocol?
(a) ATA over SCSI (b) IP over SCSI
(c) SCSI over ATA (d) SCSI over IP

11. Many FCAL nodes are attempting to simultaneously send I/O to a storage array. How is control of the loop determined?
(a) Hub bypass circuit (b) Hub name service
(c) Node that owns the token (d) Node with the highest priority

12. What is the role of a bus?
(a) Control point for all host activities and resources
(b) Device for storing data
(c) High speed interconnect between CPU, memory, and disks
(d) Temporary location for commands and data

13. A LUN is created from three disks using one partition from each of the disks. When this LUN is presented to the host, how does the host view this LUN?
(a) As three separate partitions (b) As a striped partition
(c) As a concatenated partition (d) As a single disk

14. Which statement accurately describes parity-based RAID?
   (a) Parity is read each time a read is performed
   (b) Parity is updated only when a full stripe is read
   (c) Parity is updated each time a write is performed
   (d) Parity is updated only when writes fill a stripe

15. What is monitored as part of a SAN health check?
   (a) Disk status
   (b) Fan-in/Fan-out ratio
   (c) ISL utilization
   (d) Port status change

16. What is the role of the back-end in an intelligent disk system?
   (a) To communicate between Tag RAM and data store
   (b) To communicate between the host and the storage system
   (c) To move data between the host and the array cache
   (d) To move data between the array cache and the disks

17. What are the five core elements of a data center infrastructure?
   (a) Applications, Databases, JBOD, Networks, Storage Arrays
   (b) Applications, Databases, Servers/Operating Systems, DAS, NAS
   (c) Applications, Databases, Servers/Operating Systems, Networks, Storage Arrays
   (d) Applications, Databases, Servers/Operating Systems, SAN, JBOD

18. Which two mechanisms can be used to address threats of user privilege elevation?
   (a) Access Control Lists (ACLs) and access control to storage objects
   (b) Access control to storage objects and storage access monitoring
   (c) User authentication and Information Rights Management (IRM)
   (d) User authentication and user authorization

19. In an EMC Symmetrix array, which component provides physical connectivity to hosts?
   (a) Back-end adapter
   (b) Cache
   (c) Front-end adapter
   (d) Storage Processor

20. Which EMC Celerra component allows management of volumes and filesystems?
   (a) Control Station
   (b) Data Mover
   (c) Disk Array Enclosure
   (d) Storage Processor

ANSWERS

1. (a)	2. (b)	3. (c)	4. (b)	5. (b)
6. (a)	7. (c)	8. (c)	9. (c)	10. (d)
11. (d)	12. (c)	13. (c)	14. (c)	15. (d)
16. (c)	17. (b)	18. (d)	19. (a)	20. (d)

INDEX

INDEX

A

B

C

D

E

S

T

U

V

W

Win Prizes

ATTENTION READERS

Win Prizes

We request for your frank assessment regarding various aspects of the book as given below :

INFORMATION STORAGE AND MANAGEMENT

By: PANKAJ SHARMA

Please Fill up the Sheet in Neat and Clean Writing

(i) What topics of your syllabus, important for your examination point of view, are not covered in this book ?

....

....

....

....

(ii) What are the chapters wherein the treatment of the subject matter is not systematic or organised or updated ?

....

....

....

....

(iii) Have you come across misprints/mistakes/factual inaccuracies in the book? Please specify the chapters and the page numbers.

....

....

....

....

(iv) What other books on the same subject have you found/heard better than the present book? Please specify in terms of price and quality

....

....

....

....

(v) Further suggestions and comments for the improvement of this book.

....

....

....

....

(PS: If need be please attach a separate sheet to write your views)

OTHER DETAILS

(i) Who recommended you this book ? (Please tick (✓) in the box near the option relevant to you.

☐ Teacher ☐ Friends ☐ Bookseller

(ii) Name of the recommending teacher, his designation and address :

(iii) Name and address of the bookseller you purchased the book from :

(iv) Name and address of your institution (Please mention the University or Board as the case may be)

(v) Your name and complete postal address :

(vi) Course you are enrolled in :

The best assessment will be awarded each month: The award will be in the form of our publications as decided by the Editorial Board.

Please mail the filled up coupon at your earliest to :

S.K.KATARIA & SONS®
4760-61/23, Ansari Road,
Daryaganj, New Delhi-110002
☎: +91-11-23243489, 23269384
E-mail : katariabooks@yahoo.com
katariabook@gmail.com